MIX
Papier aus verantwortungsvollen Quellen
Paper from responsible sources
FSC® C105338

Dr. Gebhu Ndlovu

Growth of Antimony on Copper

A Scanning Tunneling Microscopy Study

Anchor Academic
Publishing

Ndlovu, Gebhu: Growth of Antimony on Copper. A Scanning Tunneling Microscopy Study, Hamburg, Anchor Academic Publishing 2017

Buch-ISBN: 978-3-96067-162-6
PDF-eBook-ISBN: 978-3-96067-662-1
Druck/Herstellung: Anchor Academic Publishing, Hamburg, 2017

Bibliografische Information der Deutschen Nationalbibliothek:
Die Deutsche Nationalbibliothek verzeichnet diese Publikation in der Deutschen Nationalbibliografie; detaillierte bibliografische Daten sind im Internet über http://dnb.d-nb.de abrufbar.

Bibliographical Information of the German National Library:
The German National Library lists this publication in the German National Bibliography. Detailed bibliographic data can be found at: http://dnb.d-nb.de

All rights reserved. This publication may not be reproduced, stored in a retrieval system or transmitted, in any form or by any means, electronic, mechanical, photocopying, recording or otherwise, without the prior permission of the publishers.

Das Werk einschließlich aller seiner Teile ist urheberrechtlich geschützt. Jede Verwertung außerhalb der Grenzen des Urheberrechtsgesetzes ist ohne Zustimmung des Verlages unzulässig und strafbar. Dies gilt insbesondere für Vervielfältigungen, Übersetzungen, Mikroverfilmungen und die Einspeicherung und Bearbeitung in elektronischen Systemen.

Die Wiedergabe von Gebrauchsnamen, Handelsnamen, Warenbezeichnungen usw. in diesem Werk berechtigt auch ohne besondere Kennzeichnung nicht zu der Annahme, dass solche Namen im Sinne der Warenzeichen- und Markenschutz-Gesetzgebung als frei zu betrachten wären und daher von jedermann benutzt werden dürften.

Die Informationen in diesem Werk wurden mit Sorgfalt erarbeitet. Dennoch können Fehler nicht vollständig ausgeschlossen werden und die Diplomica Verlag GmbH, die Autoren oder Übersetzer übernehmen keine juristische Verantwortung oder irgendeine Haftung für evtl. verbliebene fehlerhafte Angaben und deren Folgen.

Alle Rechte vorbehalten

© Anchor Academic Publishing, Imprint der Diplomica Verlag GmbH
Hermannstal 119k, 22119 Hamburg
http://www.diplomica-verlag.de, Hamburg 2017
Printed in Germany

ACKNOWLEDGMENTS

I would like to express my deep and sincere gratitude to both my promoters Prof. Wiets D. Roos and Prof. Thembela K. Hillie for their assistance throughout the project, and always making time to discuss the endless list of questions, for some very useful comments regarding presentation and interpretation of the results presented in this thesis. They made the project a success and I am humbled by their intellect. Their wide knowledge, logical way of thinking, understanding nature, constant encouragement and guidance have provided a good basis for the thesis. I am grateful to Prof Hendrik Swart and the Physics department as a whole at the University of the Free State for their understanding and patience.

I am grateful and to Drs. Joseph Asante and Bonex Mwakikunga for their observations and comments which helped in establishing the overall direction of the research. My sincere thanks to the entire National Centre for Nano–Structured Materials (CSIR) academic and administrative staff for all the assistance and love we shared from my arrival at the centre, especially Simphiwe Mcineka and Margaret Ward.

I would like to express my gratitude to Prof. Thembela Hillie together with the collaboration with Prof Martin Ntwaeaborwa for arranging the visit to KIST which is where I learned a great deal about UHV systems and growth of thin films.

I am grateful to the staff at KIST, Centre for Spintronics Research, especially Drs. Jin Ding Song, Hyung–Jun Kim, Kyung Ho, and Hyung Cheol Koo for teaching me the experimental techniques involved in growing, characterizing and fabrication of spin devices using MBE, and Sdh measurement systems. I thank the members of the then LDS group (Drs. Gugu Mhlongo, Simon Dhlamini and Mr. Thomas Malwela) for valuable discussions and motivation during both good and hard times. Many thanks goes to Matete Mashapa for his assistance on the modeling side of things and Charl Jafta for his valuable assistance regarding segregation studies. I am grateful to the CSIR, Department of Science and Technology (DST) and the NRF, for the research scholarship, grants to conferences and meetings and funding the project as a whole.

I want to express my deepest sense of gratitude to my mother Duduzile S. Mbatha, brothers (Sbu, Mxo, Mcliff and Toto) and sisters (Busi and Hloni) for tolerating my absence in their lives for all the years I spent away from home and for their everlasting love and support. My daughter Siphosazo, nephews Lunga and Juniour and friends were the source of endless inspiration and constant support during my PhD; big thanks to all of you and I love you always. Finally heart full thanks to Mantapo Ramotsabi for listening to all my troubles and helping me with everything and standing by my side always. She made this possible I would not have done it without her support. Last but not the least; I would like to thank almighty God for giving me strength and courage to complete this work, He is my pillar of strength always and I will dwell in his house.

ABSTRACT

The thesis deals with adsorption, self–assembly and surface reactions of Sb atoms on solid Cu(111) substrates. It is of genuine interest in materials science and technology to develop strategies and methods for reproducible growth of extended atomic and molecular assemblies with specific and desired chemical, physical and functional properties. When the mechanisms controlling the self-organized phenomena are fully disclosed, the self-organized growth processes can be steered to create a wide range of surface nanostructures from metallic, semiconducting and molecular materials. The experimental technique used to study ordered phases and phase transitions of Sb on Cu(111) substrates was the Scanning Tunneling Microscopy (STM) – an outstanding method to gain real space information of the atomic scale realm of adsorbates on crystalline surfaces. It is a general trend to conduct studies on well-known structures before one begins working on complicated systems. Therefore, in this study, Si(111) Cu(111) and HOPG surfaces were studied in atomic detail to confirm the calibration and the resolution capabilities of the VT STM system. The acquired data from STM LEED studies were comparable to the reported theoretical and experimental data in literature.

The investigated Cu(111) – Sb system is characterized by a complex interplay between adsorbate interactions and adsorbate substrate interactions which in this study manifests through self–assembly processes. Both LEED and AES were utilized to determine the substrate cleanliness prior to the growth of monolayer Sb coverage (0.43 ± 0.02 ML Sb as calculated from the acquired STM data). The freely diffusing Sb adatoms on the copper surface were thermally excited from a random distribution of Sb atoms after growth to a finally rearrangement to more energetically stable configuration. The experimental results illustrated the presence of a surface alloy after annealing at ~360°C. The Cu – Cu spacing increased from 0.257 ± 0.01 nm (atomically clean Cu(111)) to 0.587 ± 0.02 nm after annealing at 360°C. At that temperature, the STM images showed the surface protrusions of different sizes and contrast, attributed to Cu and Sb atoms.

In addition to the conventional $(\sqrt{3} \times \sqrt{3})R30°$–Sb structural phase acquired at ~400°C, new metastable structural phases: $(2\sqrt{3} \times 2\sqrt{3})R30°$–Sb stable up to 250°C and $(2\sqrt{3} \times \sqrt{3})R30°$–Sb stable up to 350°C were obtained for the first time after annealing at 600°C and 700°C, respectively. The $(\sqrt{3} \times \sqrt{3})R30°$–Sb phase at 400°C showed an increase in atomic spacing between Cu atoms (0.626 ± 0.001 nm) as compared to the surface alloy at 300°C (Cu-Cu spacing 0.460 ± 0.002 nm) and the asymmetry at the sample surface was clearly evident. High resolution LEED studies revealed a rich phase showing extra spots on the commensurate $(\sqrt{3} \times \sqrt{3})R30°$–Sb phase compared to the clean Cu(111) substrate. STM data after annealing at 600°C and 700°C was best described by a structural model involving an ordered p(2×2) and p(2×1) overlayer structures superimposed onto the $(\sqrt{3} \times \sqrt{3})R30°$–Sb surface, respectively. At elevated temperatures LEED showed ring shaped diffraction patterns composed of twelve equidistant spots which are consistent with the growth of a hexagonal film forming three equivalent rotational domains. All the superstructures were found to favour a structural model based on Sb atoms occupying substitutional rather than overlayer sites within the top Cu(111) layer.

Other than the dissolution of Sb onto Cu(111), the study report also on the segregation of Sb on Cu together with STS measurements. The surface chemical reactivity on an atom–by–atom basis of the Cu sample surface was studied by current imaging tunneling spectroscopy (CITS). The local density of states (LDOS) were derived from dI/dV maps at low tunneling voltages by a simultaneous measurement of high resolution topographic micrographs. The onset of the atomically clean Cu(111) surface was determined to be ~8 meV below E_F which reflects the onset of tunneling from the surface state band as determined by scanning tunneling spectroscopy. The use of surface sensitive techniques (LEED, AES, STM, STS) in studying the surface alloy in question has enabled more precise statements to be made about the surface structure of the system at various temperatures. Based on the experimental results, a comprehensive study of the adsorption and segregation behavior of Sb on Cu(111), including the mechanisms for phase formation at the atomic scale is presented in this study.

KEY WORDS

Self–assembly, Antimony, Copper, Organized growth, Scanning tunneling microscopy and spectroscopy, Adsorbates, Monolayers, Diffusion, Superstructure, Surface alloys, Surface tension, Root three by root three, Surface sensitive, Nanostructures, Annealing, Surfactant, Silicone, Highly oriented pyrolytic graphite, Segregation, Activation energy, Tip fabrication, Surface reconstruction, Structural phases, Density of states.

TABLE OF CONTENTS

CHAPTER 1
INTRODUCTION 1
1.1 BACKGROUND 1
1.2 OBJECTIVES OF PRESENT WORK 8
1.3 THESIS OUTLINE 9
1.4 REFERENCES 11

CHAPTER 2
THEORY 19
2.1 THE FCC(111) SURFACE STRUCTURE 20
2.2 RELAXATION AND RECONSTRUCTION OF SURFACES 21
    2.2.1 RELAXATION 22
    2.2.2 RECONSTRUCTION OF CLEAN SURFACES 22
    2.2.3 ADSORBATE–INDUCED RECONSTRUCTION 23
2.3 ADSORPTION 24
2.4 KINETICS OF ADSORPTION 25
2.5 THE $\sqrt{3}$ STRUCTURES ON (111) FCC METAL SURFACES 26
2.6 DYNAMICS OF ADATOMS DIFFUSION 27
2.7 INTERACTION AT THE SURFACE 29
    2.7.1 INFLUENCE OF ADSORBATES ON SURFACES 29
2.8 TYPES OF GROWTH 31
    2.8.1 FRANK–VAN DER MERWE (FM) OR LAYER–BY LAYER 32
    2.8.2 VOLMER–WEBER (VW) OR 3D ISLAND 33
    2.8.3 STRANSKI–KRASTANOW (SK) OR 3D ISLAND–ON WETTING–LAYER GROWTH 33
2.9 INTRODUCTION TO SEGREGATION 33
2.10 THE MODIFIED DARKEN MODEL 34
2.11 REFERENCES 36

CHAPTER 3

SCANNING TUNNELING MICROSCOPY (STM) 40
3.1 INTRODUCTION 40
3.2 ELECTRON TUNNELING 40
3.3 STM PRINCIPLE OF OPERATION 42
3.4 TUNNELING CURRENT 44
3.5 BARDEEN THEORY OF TUNNELING 46
3.6 MODES OF OPERATION 48
    3.6.1 CONSTANT CURRENT IMAGING MODE 48
    3.6.2 CONSTANT HEIGHT IMAGING MODE 50
3.7 SCANNING TUNNELING SPECTROSCOPY (STS) 51
    3.7.1 CURRENT IMAGING TUNNELING SPECTROSCOPY (CITS) 52
    3.7.2 CONSTANT CURRENT SPECTROSCOPY 53
    3.7.3 CONSTANT VOLTAGE SPECTROSCOPY 53
3.8 STM MAJOR COMPONENTS 54
    3.8.1 VIBRATION ISOLATION 54
    3.8.2 ELECTRONIC FEEDBACK CONTROL SYSTEM 55
    3.8.3 THE STM TIP 55
    3.8.4 IMAGES AND FILTERING 56
3.9 ATOMIC RESOLUTION ON WELL KNOWN SUBSTRATES 57
3.10 HIGHLY ORIENTED PYROLYTIC GRAPHITE (HOPG) 58
    3.10.1 EXPERIMENTAL 59
    3.10.2 RESULTS AND DISCUSSION 60
3.11 SI(111) – 7×7 RECONSTRUCTION 64
    3.11.1 EXPERIMENTAL 65
    3.11.2 RESULTS AND DISCUSSION 66
3.12 REFERENCES 69

CHAPTER 4

EXPERIMENTAL FACILITIES AND PROCEDURES 73
4.1 TIP FABRICATION 73
4.2 CHARACTERIZATION TECHNIQUES 76
    4.2.1 THE UHV VARIABLE TEMPERATURE STM SYSTEM 76

    4.2.2   LOW–ENERGY ELECTRON DIFFRACTION (LEED) 79

    4.2.3   AUGER ELECTRON SPECTROSCOPY (AES) 81

  4.3   REFERENCES 84

CHAPTER 5

DISSOLUTION OF SB ON CU(111) 85

  5.1   INTRODUCTION 85

  5.2   EXPERIMENTAL PROCEDURE 87

  5.3   RESULTS AND DISCUSSION 89

    5.3.1   ATOMICALLY CLEAN CU(111)... 89

  5.4   GROWTH OF SB ON CLEAN CU(111) SURFACES 91

  5.5   STAGES OF DISSOLUTION 95

    5.5.1   ANNEALING AT 360°C 95

    5.5.2   ANNEALING AT 400°C 97

    5.5.3   ANNEALING AT 600°C 103

    5.5.4   ANNEALING AT 700°C 109

  5.6   SEGREGATION OF SB ON CU(111)... 112

  5.7   SUMMARY 114

  5.8   REFERENCES 116

CHAPTER 6

SCANNING TUNNELING SPECTROSCOPY (STS)... 118

  6.1   INTRODUCTION 118

  6.2   STS OF CU(111) AT ROOM TEMPERATURE 120

  6.3   SUMMARY 126

  6.4   REFERENCES 127

CHAPTER 7

SUMMARY AND CONCLUSION 128

FUTURE PROSPECTS 134

PUBLICATIONS... 135

INTERNATIONAL MEETINGS AND CONFERENCES 136

LOCAL MEETINGS AND CONFERENCES 136

AWARDS 138

ACCRONYMS

Sb	Antimony	
Ag	Silver	
Cu	Copper	
2–DEG	Two–dimensional electron gas	
AES	Auger electron spectroscopy	
HOPG	Highly oriented pyrolytic graphite	
LEED	Low energy electron diffraction	
LDOS	Local density of states	
STM	Scanning tunneling microscopy	
STS	Scanning tunnelling spectroscopy	
UHV	Ultra–high vacuum	
VT–STM	Variable temperature scanning tunneling microscopy	
E_F	Fermi energy	
FCC	Face centered cubic	
HCP	Hexagonal closed packed	
3D	Three dimensional	
RT	Room temperature	
BZ	Brillouin zone	
FHUC	Faulted half unit cell	
UHUC	Unfaulted half unit cell	

CHAPTER 1

INTRODUCTION

1.1 Background

Materials and their properties have fascinated the human race since the past centuries when fabrication was regarded more as an art than a science [1–3]. Today's way of life depends heavily on the ability of a few million transistors to process data and a few million more to store it, whether temporarily or permanently [4–10]. Until now, investigative research is still carried out in the field of materials and surface science which is primarily intended to expand the range of knowledge and properties of materials of various types for both technological and fundamental applications [4–12]. Given that most material products come in the form of alloys, some of the ongoing questions in the field of materials science includes, whether different elemental materials are immiscible when combined under controlled conditions, whether or not the materials dissolve in one another, and, if they do, to what extent. Alternatively will they react to form a compound? If so, in what atomic ratios (or stoichiometry)? And how does processing conditions influence the product of the combined materials?

The field of modern surface science is driven by the ever increasingly specialized intricate set of characterization tools that are drawn from a wide range of scientific disciplines utilized to give answers to some, if not all of the above–mentioned questions with a high degree of accuracy [11–17]. Due to the demand for faster processing speeds and huge storage capacity of electronic devices, the study of materials has ushered in the field of nanoscience and nanotechnology where materials are investigated in dimensions less than 100 nm [4,5,18–20]. In particular, the control of the shape and size in nanostructural growth has been a very hot research field topic [19–21], since the properties and applications of nanostructures mainly depend on their shape, size and surface morphology [9, 22–26].

It can be argued with considerable justification that the field of nanotechnology is driven by the possibility of fabricating tailor–designed nanostructures with unique properties. Two different approaches are generally used in the fabrication of nanostructures, namely, top–down and bottom–up. The top–down method mainly includes photolithography and etching techniques which permit the creation of nanostructures over large sample areas [4,5,18]. Other examples of top–down include ball milling and arc discharge. The photolithography process has some disadvantages, such as the sizes of nanostructures are limited by wavelength of the photons used in the photolithography and mask sizes [4,5,20]. On the other hand, in bottom–up approach the building block materials for fabricating self–assembled nanostructures are atoms, molecules or clusters [4–7,18,20]. Bottom–up examples include chemical vapour deposition (CVD) [27,28], molecular beam epitaxy (MBE) [29-31], sputtering, liquid to solid nucleation, pulsed laser deposition (PLD) [32–34], sol–gel techniques [35–37], to mention but a few. The self-assembled nanostructures can be formed in a growth environment taking advantages of some energetic, geometric and kinetic effects of over–layer materials and substrates where molecules diffuse from atomic to the mesoscopic scale [18,20,38]. In this study, the latter approach is utilized to grow thin film layers by controlling the flux of the deposited material, the kinetics and thermodynamics of the underlying substrate.

Crystalline surfaces of noble metals (Cu, Ag, Au) have broken translational bulk symmetry [39]. The lack of translational symmetry along the surface normal results in electronic states localized perpendicular to the surface, the so-called surface states [40–42]. These surface states are localized to the first few atomic layers at the surface and form a quasi two-dimensional free electron gas (2DEG) which is confined to the first few atomic layers at the crystal surface [6,43,44]. Therefore, the electrons strongly influence the interaction between the surface and its environment and thereby contribute to the chemistry of surfaces, such as adsorption processes, equilibrium surface structures, or catalytic reactions [10,54,46]. Furthermore, surface states are responsible for long-range substrate mediated adsorbate interactions, which dominate the bulk-state mediated contribution for large adsorbate-adsorbate separation [47,48]. In addition, the contribution from surface states is relevant for the total energy balance of surface reconstructions [6,40].

Thin films are of special interest especially when they have a thickness of only a few monolayers. Properties of such films may drastically differ from their bulk due to the surface restriction to two dimensions and due to the interaction with the substrate [49–51]. The surfaces of bulk alloys are of practical interest for their chemical properties be it novel activity or selectivity to certain reactions [52–54] in a way which differs from the constituent elements in isolation or novel passiveness to corrosion. Some of the basic thermodynamics of segregation in alloys is far from new. Nevertheless, our understanding of these chemical and physical phenomena is far from complete, and the application of surface science methods to investigate these phenomena is a manifestation of a general trend to the study of surfaces of increasing complexity.

Copper (Cu) has proved to be a favoured substrate for studies of ultra–thin metallic film growth for many reasons, including the relative ease of cleaning and maintaining surface cleanliness, the high level of crystalline quality and the advantages of a full d–band electronic configuration allowing high resolution studies of the surface and bulk electronic structure [55]. Some surface adsorbate species such as antimony (Sb) and indium (In) are known to act as surfactants in both homo – and hetero – epitaxy because of their low surface energy [56,57]. These adsorbates appear to induce layer–by–layer growth [58] in systems which otherwise tend to island growth, whilst remaining at the surface as the growth proceeds rather than being incorporated into the growing film [59,60]. Despite increased interest in the applications of surfactants on metal growth, the microscopic mechanisms of dissolution and segregation of Sb is not yet fully understood [61]. Established examples of this phenomenon is the role of Sb as a surfactant on the growth of Co on Cu(111) [62] and the growth of Ag on Ag(111) [63,64].

The phenomenon of surface segregation is defined as the preferential enrichment of one component of a multi–component system at a boundary or interface [65–68]. The extent of segregation is influenced by strain energy due to the atomic size mismatch between the solute and the solvent, as well as the differences in their surface energies [52,69,70]. During surface segregation the surface enthalpy is different from the bulk enthalpy and occurs at finite temperatures (or in the materials growth process) when energy activation barriers to diffusion are overcome.

When monolayers of Sb are grown on Cu(111) or Ag(111) there are generally two proposed atomic structural models [71,72]. On the first model, Sb atoms are located above the (111) surface in face centered stacking sites as adatoms, while the second structure, involves Sb atoms sitting substitutionally within the (111) surface layer of the substrate [71–74]. Previous experimental [71–73] and theoretical [74,75] studies have suggested that the energetics of this systems are such that in the ordered 0.33 ML Ag(111) $(\sqrt{3} \times \sqrt{3})R30°$–Sb or Cu(111) $(\sqrt{3} \times \sqrt{3})R30°$–Sb phase, the Sb atoms substitute one–third of the outermost substrate (Ag or Cu) atoms to produce an ordered surface alloy which shows out–ward relaxation of the Sb atoms. Thus these previously reported studies rules out the first proposition of Sb sitting as adatoms on the substrate.

The formation of the $(\sqrt{3} \times \sqrt{3})R30°$ superstructure, is well understood in the case of Sb–Ag system since Ag and Sb form bulk intermetallic compounds and their atomic radii mismatch is small (~1%), however, it is rather difficult to understand how such a phase can form in the Sb–Cu system because of the large atomic mismatch (~ 15%) between Cu and Sb [76,77]. Therefore, the observation of almost identical two–dimensional surface alloys in both cases reveals that the main driving forces for formation is the tendency to chemical ordering, almost independently of size–mismatch between deposit and substrate atoms. The Cu – Sb system has been widely studied for (100) and (111) Cu surface orientations [76,77], considering both segregation and dissolution of deposited Sb layers. Previous *in situ* STM, Auger electron spectroscopy (AES), and low energy electron diffraction (LEED) studies of the dissolution and segregation of ~ 1 ML of Sb on Cu(111) has revealed that at ~ 400°C, the dissolution stops, due to formation of a surface alloy which exhibits a p$(\sqrt{3} \times \sqrt{3})R30°$ superstructure which is fully consistent with one Sb atom per unit cell in the p$(\sqrt{3} \times \sqrt{3})R30°$ structure [78]. The dissolution and segregation kinetics are thus closely linked to the equilibrium surface segregation.

At higher Sb coverages, a p(2×2) reconstruction has been observed on Ag(111), and explained as an ordered p(2×2)–Sb overlayer superimposed on the $(\sqrt{3} \times \sqrt{3})R30°$–Sb surface [72,73]. A similar structure at higher Sb coverages has not been reported on the CuSb system. Theoretical calculations have shown that the creation of $(\sqrt{3} \times \sqrt{3})R30°$–Sb surface alloy is energetically favoured [75]. The mechanism underlying the dissolution and segregation of the large Sb (atomic radius = 0.159 nm) atoms deposited on the (111) close–packed Cu (atomic radius = 0.128 nm) surface is still an open topic for investigation.

From an experimental point of view, the first available methods capable of investigating surfaces at the atomic level were diffraction methods in the late 20s [17,79] followed by scanning tunneling microscopes in the late 70s [80–82]. Quantifying and understanding the structure of surfaces, and particularly of adsorbates on surfaces, is a key step to understanding many aspects of the behavior of surfaces including the electronic structure and the associated chemical properties. Both low–energy electron diffraction (LEED) and reflection high–energy electron diffraction (RHEED) has for example brought many successes in understanding atomic surface reconstruction [83–85]. One of the major disadvantages of these methods is that the surface is represented in the reciprocal space, making interpretation of measured physical data more intricate.

The first real space experimental technique capable of imaging atoms at a surface is the field ion microscope (FIM) [86–88]. While this method has been widely and successfully used to study the diffusion of adatoms and clusters at a surface, it is not suited to investigate large areas of nanostructure–covered singular surfaces.

The cornerstone of modern growth investigation was laid by the invention of the scanning tunneling microscope (STM) in 1982 by Binnig and Rohrer (Nobel laureates in 1986) [80,81] which will be the main experimental tool for this study. The invention of the STM solved one of the most intriguing problems which fascinated surface scientist for quite some time, the Si(111)–(7×7) reconstruction [81,89]. Other than providing images of surfaces and adsorbate atoms and molecules with unprecedented resolution, the STM has also been used previously to modify surfaces, for example by locally pinning molecules to a surface [90] and by transfer of an atom from the STM tip to the surface [91]. Recent developments in nanoscience make it possible to engineer artificial structures at surfaces [92–99]. The STM tip can be used as an engineering or analytical tool, to fabricate artificial atomic–scale structures where novel quantum phenomena can be probed and properties of single atoms and molecules can be studied at an atomic level [82,100,101]. These structures include quantum corrals [100,101] and atomic chains [102].

The STM can also be operated in the spectroscopy mode (STS) to study the local electronic structure of a sample's surface. This is usually done by sweeping the bias voltage V and measuring the tunneling current I while maintaining constant tip–sample separation z which results in current vs. voltage (I–V) curves characteristic of the electronic structure at a specific x,y location on the sample surface [103-105]. By numerically differentiating I–V, the conductance dI/dV can be obtained. The interpretation of dI/dV spectra can be complex but it can be shown that, under ideal conditions, dI/dV is a good measure of the sample density of states (DOS) [103-105]. This method makes it possible to get energy spectra of very small objects at the surface. This can be used in the microelectronics industry, for example, to control electrical properties of transistors, especially those of a very small size.

1.2 Objectives of present work

The main aim of this project is to utilize an Ultra–High Vacuum Variable Temperature Scanning Tunneling Microscopy (UHV VT–STM) to study the dissolution and segregation of fractional monolayer of antimony grown on copper (111) surfaces *in–situ*.

The main objectives are:

(1) Optimize the VT–STM system in order to obtain morphology of Cu(111), HOPG and Si(111) with atomic scale resolution in order to calibrate the STM especially using the well-known Si surfaces

(2) Systematic study the nature of surface reconstruction of conducting and semiconducting material in UHV

(3) Investigate the atomic interaction of the Cu–Sb system and its various structural phases as a function of annealing temperature

(4) Obtain local Auger spectral data at various temperatures on the segregation of Sb to the Cu(111) surface in order to calculate diffusivity, segregation energy and activation energy in such a system at the nano–scale and compare their bulk values available in literature

(5) Perform current–voltage measurements to obtain valuable information regarding the electronic structure of the Cu–Sb system

1.3 Thesis outline

The thesis is structured as follows:

Chapter 2, introduces the theoretical background on surfaces and various processes such as reconstructions at surfaces, the kinetics of thin film growth and types of growth. The chapter also introduces the basic concepts of segregation and the modified Darken model which describes segregation phenomenon for binary systems.

Chapter 3 focuses on the main experimental method of the thesis, the scanning tunneling microscopy (STM). The chapter contains the necessary background information to understand the experimental work presented later in the thesis. These include the theory of electron tunneling and the operation principles of the STM system. The theoretical background on scanning tunneling spectroscopy which forms part of the STM is also explained in detail in this chapter. The chapter also explores the unprecedented resolution of the STM by studying the surfaces of well-known substrates such as Si(111) and HOPG in detail.

Chapter 4 comprises the experimental techniques and procedures. These include the theory of surface sensitive techniques (AES, LEED). The UHV VT–STM system and its components together with tip fabrication processes are also discussed.

Chapter 5 looks at the adsorption and dissolution studies of sub–monolayer Sb grown on Cu(111) surface to get better insight into the atomic structure/s of the surface alloy and the influence of temperature at the sample surface. In addition, the chapter reports on the experimental segregation studies of Sb on the copper sample.

Chapter 6 looks at the added advantage of being able to acquire electronic information simultaneously with topographical data utilizing STS which allow direct comparisons of the topography and the electrical characteristics of the sample surface.

Chapter 7 focuses on discussions and conclusions. Key points regarding experimental results from dissolution, segregation and the conductance measurements are summarized to give a better insight with a complete conclusion of the study.

1.4 References

[1] V. Biringuccio (in 16th Century Italian). *The Pirotechnia of Vanoccio Biringuccio*. *Dover Publications*. ISBN 0486261344. 20th Century translation by C. S. Smith and M. T. Gnudi, (1990)

[2] A. Silcock, and M. Ayrton (reprinted 2003). *Wrought iron and its decorative use*: with 241 illustrations. Mineola, N.Y: Dover. pp. 4. ISBN 0–486–42326–3. See also, 1570 Chapter XXVIII, The Autobiography of Benvenuto Cellini, as translated by John Addington Symonds, Dolphin Books edition, (1961)

[3] R. H. Fowler, and L. Nordheim, Proc. Roy. Soc. A 11, 173 (1928) 719

[4] C. Dupas, P. Houdy, and M. Lahmani, *Nanoscience, Nanotechnologies and Nanophysics*, Springer, (2006)

[5] W. R. Fahrner, *Nanotechnology and Nanoelectronics, Materials, Devices, Measurement Techniques*, Springer, (2005)

[6] S. G. Davison, and M. Steslicka: *Basic Theory of Surface States*, Oxford University Press, New York, (1992)

[7] M. Meixner, Ph.D. thesis, Institut für Theoretische Physik, Technische Universität Berlin (2002)

[8] B. H. Hamadani, PhD thesis, Rice University, (2007)

[9] H. Ulbricht, Surf. Sci. 603 (2009) 1853–1862

[10] G. A. Somorjai, and J.Y. Park, Catalysis Letters 115 (2007) 87

[11] P. C. Thüne, and J. W. Niemantsverdriet, Surf. Sci., 603 (2009) 1756–1762

[12] G. A. Somorjai, and J. Y. Park, Surf. Sci., 603 (2009) 1293–1300

[13] M. A. Paesler, and P. J. Moyer, Near–Field Optics: *Theory, Instrumentation, and Applications*, New York: Wieley–Interscience (1996)

[14] A. Lafosse, M. Bertin, A. Hoffman, and R. Azria, Surf. Sci. 603 (2009) 1873–1877

[15] A. Jablonski, Surf. Sci. 603 (2009) 1342–1352

[16] S. Carrara, A. Cavallini, Y. Maruyama, E. Charbon, and G. De Micheli, Surf. Sci. 604 (2010) L71–L74

[17] W. L. Bragg, *The analysis of crystals by the X–ray spectrometer*, Proc. R. Soc. Lond. A 89 (1914) 613

[18] J. V. Barth, and G. Costantini, K. Kern, Nature, 437 (2005)

[19] B. Panigrahy, M. Aslam, D. S. Misra, and D. Bahadur, Cryst Eng Comm, 11 (2009) 1920–1925

[20] G. Zhang, K. Tateno, H. Gotoh, and T. Sogawa, NTT Technical Review, 8 (2010) 8

[21] M. Cao, C. Guo, Y. Qi, C. Hu, and E. Wang, J. Nanosci Nanotech. 4, 7 (2004) 29–32

[22] J. Wintterlin, and M. L. Bocquet, Surf. Sci. 603 (2009) 1841–1852

[23] M. Zinke–Allmang, L. C. Feldman, and M. H. Grabow, Surf. Sci. Rep. 16, 377 (1992)

[24] R. Wiesendanger, M. Bode, M. Kleiber, M. Lohndorf, R. Pascal, and A. Wadas. D. Weiss, J. Vac. Sci. Technol. B 15, 4 (1997)

[25] N. A. Khan, A. Uhl, S. Shaikhutdinov, and H.–J. Freund, Surf. Sci. 600 (2006) 1849–1853

[26] W. Ki Hong, B. Joong Kim, T. Wook Kim, G. Jo, S. Songa, S. Shin Kwon, A.Yoon, E. A. Stach, and T. Lee, Colloids and Surfaces A: Physicochem. Eng. Aspects 313–314 (2008) 378–382

[27] P. John, *The Chemistry of the Semiconductor Industry*, Chapman and Hall: New York (1987)

[28] H. O. Pierson, *Handbook of Chemical Vapor Deposition (CVD) – Principles, Technology and Applications* (2nd Edition) William Andrew Publishing/Noyes (1999)

[29] P. Velling, W. Fix, W. Geibelbrecht, W. Prost, G.H. Dohler, F. J. Tegude, Journal of crystal growth 195 (1998) 490–494

[30] R. F. C. Farrow, *Molecular Beam Epitaxy, Applications to Key Materials*, William Andrew Publishing (2001)

[31] S. Thainoi, S. Suraprapapich, M. Sawadsaringkarn, and S. Panyakeow, Solar Energy Materials and Solar Cells, 90,18–19 (2006) 2989–2994

[32] R. Eason, *Pulsed laser deposition of thin films*, Wiley–Interscience (2007)

[33] D. B. Chrisey, and G. K. Hubler, *Pulsed laser deposition of thin films*, J. Wiley (1994)

[34] R. Savua, and E. Joanni, Scripta Materialia 55 (2006) 979–981

[35] Y. L. Zub, and V. G. Kessler, *Sol–Gel Methods for Materials Processing*, Springer (2008)

[36] L. H. Lee, O. Gunawan, B. S. Ooi, Y. Zhou, Y. C. Chan, and Yee Loy, Lam Proc. SPIE 3899 (1999) 162

[37] M. N. Uddin, and Y. S. Yang, J. Mater. Chem. 19 (2009) 2909–2911

[38] J. A. Liddlea, Y. Cui, and P. Alivisatos, J. Vac. Sci. Technol. B 22 (2004) 6

[39] L. Petersen, and P. Hedegard, Surf. Sci. 459 (2000) 49–56

[40] B. Lazarovits, L. Szunyogh, and P. Weinberger, Phys. Rev. B 73 (2006) 045430

[41] O. Sánchez, J. M. García, P. Segovia, J. Alvarez, A. L. Vázquez de Parga, J. E. Ortega, M. Prietsch, and R. Miranda , Phys. Rev. B 52 (1995) 7894–7897

[42] L. Bürgi, N.Knorr, H. Brune, M. A. Schneider, and K. Kern, Appl. Phys. A 75 (2002)

[43] P. M. Echenique, and J. B. Pendry, Prog. Surf. Sci. 32 (1990) 111

[44] R. Matzdorf, Surf. Sci. Rep. 30 (1998) 153

[45] J. W. Niemantsverdriet, *Spectroscopy in Catalysis*, VCH Publishers (1995)

[46] A. J. Bennett, Phys. Rev. B 1 (1970) 203–207

[47] S. Zhang, E. R. Hemesath, D. E. Perea, E. Wijaya, J. L. Lensch–Falk, and L. J. Lauhon, Nano Lett. 9, 9 (2009)

[48] P. Hofmann, K. C. Rose, V. Fernandez, and A. M. Bradshaw, Phys. Rev. Lett. 75 (1995) 2039–2042

[49] R. W. Hoffman, *The mechanical properties of thin condensed films*, Physics of thin films, 3, New York, Academic Press (1966)

[50] K. L. Chopra, *Thin film phenomena*. New York: McGraw–Hill (1990)

[51] M. Murakami, T. S. Kuan, and I. A. Blech, *Treatise on materials science and technology*, 24. New York: Academic Pres (1982)

[52] P. A. Dowben and A. Miller, editors, *Surface Segregation Phenomena*, CRC Press, Boston (1990)

[53] J. J. Terblans, and G. N. van Wyk, Surf. Interface Anal. 2003; 35: 779–784

[54] K. Umezawa, H. Takaoka, S. Hirayama, S. Nakanishi, and W. M. Gibson, Current Appl. Phys. 3 (2003) 71–74

[55] D. A. King and D. P. Woodruff, *The Growth and Properties of Ultrathin Epitaxial Layers, The Chemical Physics of Solid Surfaces*, 8, Elsevier, Amsterdam (1997)

[56] H. Wider, V. Gimple, W. Evenson, and G. Schatz, J. Jaworski, J. Prokop and M. Marszalek J. Phys. Cond. Mat. 15 (2003) 1909–1919

[57] B. Aufray, H. Giordano, and D. N. Seidman, Surf. Sci. 447 (2000) 180

[58] H. A. van der Vegt, W. J. Huisman, P. B. Howes, T. S. Turner, and E. Vlieg, Surf. Sci. 365, 2 (1996) 205–211

[59] H. A. van der Vegt, H. M. van Pinxteren, M. Lohmeier, E. Vlieg and J. M. C. Thornton, Phys. Rev. Lett. 68 (1992) 3335

[60] H. A. van der Vegt, J. Alvarez, X. Torrelles, S. Ferrer and E. Vlieg, Phys. Rev. B 52 (1995) 17443

[61] J. A. Meyer, H. A. van der Vegt, J. Vrijmoeth, E. Vlieg, and R. J. Behm, Surf. Sci. 355 (1996) L375

[62] V. Scheuch, K. Potthast, B. Voigtlander, and H.P. Bonzel, Surf. Sci. 318 (1994) 115

[63] J. Vrijmoeth, H. A. van der Vegt, J. A. Meyer, E. Vlieg, and R. J. Behm, Phys. Rev. Lett. 72 (1994) 3843

[64] F. R de Boer, R. Boom, W. C. M. Mattens, A. R. Miedema, and A. K. Niessen, *Cohesion and Structure, 1,* North Holand, Amsterdam (1988)

[65] A. V. Ruban, H. L. Skriver, and J. K. Norskov, Phy. Rev. B 59 (1999) 24

[66] F. Cabané–brouty, and J. Bernardini, J. Phys. Colloques, 43, C6 (1982)

[67] P. A. Dowben, and S. J. Jenkins, *Frontiers in Magnetic Materials*, edited by A. Narlikar, Springer, Verlag (2005)

[68] P. A. Dowben, N. Wu, N. Palina, R. Müller, J. Hormes, and Ya. B. Losovyj, http://digitalcommons.unl.edu/physicsdowben/179

[69] J. J. Terblans, and G. N. vanWyk, Surf. Interface Anal. 36 (2004) 935–937

[70] H. K. Jeong, T. Komesu, C. S. Yang, P. A. Dowben, B. D. Schultz, and C. J. Palmstr, Mater. Lett. 58 (2004) 2993

[71] H. Crugel, B. Le´pine, S. Ababou, F. Solal, G. Jezequel, C.R. Natoli, and R. Belkhou, Phys. Rev. B 55 (1997)

[72] T. C. Q. Noakes, D. A. Hutt, C. F. McConville, and D. P. Woodruff, Surf. Sci. 372 (1997) 117

[73] E. A. Soares, C. Bittencourt, V. B. Nascimento, V. E. de Carvalho, C. M. C. de Castilho, C. F. McConville, A .V. de Carvalho, and D. P. Woodruff, Phys. Rev. B 61 (2000) 13983

[74] S. A. de Vries, W. J. Huisman, P. Goedtkindt, M. J. Zwanenburg, S. L. Bennett, I. K. Robinson, and E. Vlieg, Surf. Sci. 414 (1998) 159

[74] S. Oppo, V. Fiorentini, and M. Scheffler. Phys. Rev. Lett. 71 (1993) 2437

[75] D. P. Woodruff, and J. Robinson, J. Phys. Cond. Mat. 12 (2000) 7699–7704

[76] H. Giordano, O. Alem, and B. Aufray, Scr. Metall. 28 (1993) 257

[77] H. Giordano, and B. Aufrey. Surf. Sci. 307–309 (1994) 816-820

[78] I. Meunier, J. M. Gay, L. Lapena, B. Aufray, H. Oughaddou, E. Landmark, G. Falkenberg, L. Lottermoser, and R.L. Johnson. Surf. Sci. 422 (1999)

[79] W. L. Bragg, *The crystalline structure of zinc oxide*. Phil. Mag. 39, 234 (1920)

[80] G. Binnig, H. Rohrer, IBM J. Res. Develop, 44, 1 (2000)

[81] G. Binnig, H. Rohrer, Ch. Gerber, and E. Weibel, Phys. Rev. Lett. 50, 2 (1993)

[82] D. M. Eigler, and E. K. Schweizer, Nature, 34 (1990)

[83] K. Oura, V. G. Lifshits, A. A. Saranin, A. V. Zatov, and M. Katayama, *Surface science, an introduction*, Springer (2003)

[84] W. Moritz, J. Landskron, and M. Deschauer, Surf. Sci. 603 (2009) 1306–1314

[85] D. P. Woodruff, and T. A. Delchar, *Modern Techniques of Surface Science*, Cambridge, University Press (2003)

[86] T. T. Tsong, Phys. Rev. Lett. 31 (1973) 1207–1210

[87] T. T. Tsong, Physics today, 46, 24 (1993)

[88] G. Ehrlich, Physics Today, 34, 6 (1981)

[89] R. M. Tromp, R. J. Hamers, and J. E. Demu, Phys. Rev. B 34 (1986) 1388–1391

[90] J. S. Foster, J. E. Frommer, and P. C. Arnett, Nature, 331 (1998) 324–326

[91] R. S. Becker, J. A. Golovchenko, and B. S. Swartzentruber, Nature 325 (1987)

[92] G. Ertl, and H. J. Freund, Physics Today 52 (1999) 32

[93] J. Y. Park, Y. Zhang, M. Grass, T. Zhang, and G. A. Somorjai, Nano Letters 8 (2008) 673

[94] H. J. Freund, H. Kuhlenbeck, J. Libuda, G. Rupprechter, M. Baumer, and H. Hamann, Topics in Catalysis 15 (2001) 201

[95] G. A. Somorjai, K.M. Bratlie, M.O. Montano, and J.Y. Park, J. Phys. Chem. B 110 (2006) 2014

[96] A. T. Bell, Science 299 (2003) 1688

[97] C. T. Campbell, Surf. Sci. Rep. 27 (1997) 1

[98] G. A. Somorjai, J. Y. Park, Physics Today 60 (2007) 48

[99] G. Binnig, and H. Rohrer, Revs. Mod. Phys. 71, 2 (1999)

[100] S. W. Hla, J. Vac. Sci. Technol. B, 23 (2005) 1351-1360

[101] E. S. Snow, P. M. Campbell, and P. J. McMarf, Appl. Phys. Lett. 63, 6 (1993) 749-751

[102] A. Deshpande, PhD thesis, faculty of the College of Arts and Sciences of Ohio University (2007)

[103] G. Hormandinger, Phys. Rev. B. 49, 19 (1994)

[104] G. Binnig and H. Rohrer Surface Science 152–153, 1 (1985) 17–26

[105] R. M. Feenstra, Surf. Sci. 299–300 (1994) 956-979

CHAPTER 2

THEORY

The inception of modern surface science dates back to the early 1960s. The breakthrough in the field resulted from a combination of factors, including progress in ultra high vacuum technology, the development of experimental methods sensitive to the surface atomic structure and the appearance of high–speed digital computers.

A surface is created by cleaving a solid and breaking of bonds between the atoms of the cleavage planes [1–5]. The work done in forming a unit area of a new surface is called the surface energy and is denoted by γ. The surface free energy of noble metals (Cu, Ag, Au) is more than three times higher than that of van der Waals substrates like graphite. It is this surface free energy which drives adsorption and catalysis and explains why metals are very active materials in such processes [4].

It is conventional to use the z–axis for the surface normal, leaving x and y for directions in the surface. The interest in surface properties of a material usually encompasses the bulk properties of the material. Therefore, understanding of the surface properties requires a good knowledge of bulk properties. Almost all surface science studies are concerned with addition of controlled amount of foreign atoms (adsorbates) on atomically clean surfaces. There are various ways in which adsorbates or molecules can be introduced onto the host material surface (substrate) [6].

The adatoms can condense onto the surface from the vapour phase through a process called adsorption or alternatively segregate from the sample bulk, or diffuse along the surface. Knowledge of the structure of surfaces and particularly of adsorbates on surfaces is a key step to understanding many aspects of the behavior of surfaces including the electronic structure and the associated chemical properties.

2.1 The FCC(111) surface structure

Surface structures of metals, being the most studied of all surface structures illustrate the degree of reproducibility and accuracy achievable by modern surface science techniques. Many metal elements crystallize in the face–centered cubic (FCC) structure. Among them are the noble metals copper (Cu), silver (Ag), and gold (Au) [7]. In the atomic configuration, noble metals are characterized by a completely filled d–level subshell and one single s–type valence electron. In the solid state, these electrons are strongly delocalized to form an sp–derived band. Close to the Fermi–level, the electronic properties are found to be in very good agreement with that of a free–electron model [8]. Even though the surface region is in principle a three–dimensional (3D) entity having a certain thickness, all symmetry properties of the surface are 2D – meaning the surface structure is periodic only in two directions. Atoms in the bulk of FCC (111) materials have coordination number 12 while those at the surface have 9 nearest neighbours. This means that the mean–square amplitude of the surface atoms is much larger than in the bulk. The surface energies per atom increase with decreasing coordination of the surface atoms. For low index surfaces of FCC metals (figure 2.1) the stability decreases in the trend $\gamma(111) < \gamma(100) < \gamma(110)$.

The surface layer of the {111} surface has a six–fold rotation axis and three non–trivial mirror planes. The surface lattice has two different three fold hollow sites, the hexagonal close packed (HCP) site and the face centred cubic (FCC) site. The packing density is the highest for the {111} surfaces, followed by the {100} and finally the {110} surfaces.

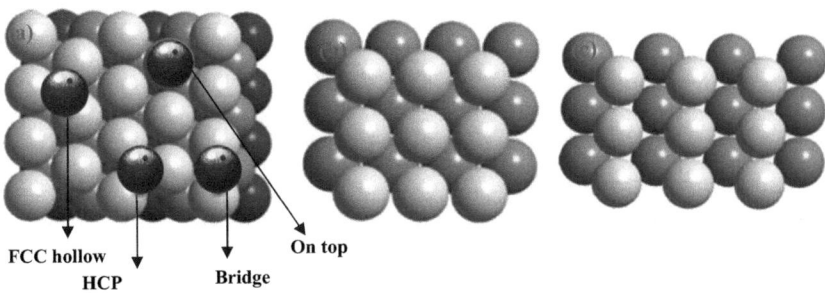

Figure 2.1. The (a) (111), (b) (100), and (c) (110) phases of a perfect FCC crystals and the adsorption sites for adatoms on the (111) FCC surface phase [9].

2.2 Relaxation and reconstruction of surfaces

The subsequent processes which involves the rearrangements of surface (and near surface) atoms to form a structure with a different periodicity and/or symmetry than that of the bulk crystal are called relaxation and reconstruction. The driving force behind this processes are the energetics of the system [4,5] (i.e. the desire to reduce the surface free energy by increasing the coordination number of surface atoms and this is achieved in several ways, as shown in figure 2.2). As with all processes, there are kinetic limitations, which prevent or hinder these rearrangements at low temperatures. Relaxations and reconstructions often occur for atomically clean surfaces in vacuum, in which the interaction with any another medium is not at play.

2.2.1 Relaxation

Atoms at the surface of a crystal have fewer neighbours than atom in the bulk. The reduction of nearest neighbour gives rise to redistribution in the selvedge. This changed force field results in the decrease in the layer spacings d perpendicular to the surface (figure 2.2 (a)) with no change either in the periodicity parallel to the surface or to the symmetry of the surface unit cell [5].

2.2.2 Reconstruction of clean surfaces

Reconstruction of surfaces is a much more readily observable effect, involving larger (yet still atomic scale) displacements of the surface atoms. Unlike relaxation, the phenomenon of reconstruction (figure 2.2 (b)) involves a change in the periodicity of the surface structure. Surface reconstruction can affect one or more layers at the surface, and can either conserve the total number of atoms in a layer (a conservative reconstruction) or have a greater or lesser number than in the bulk (a non–conservative reconstruction). The reconstructive surface phase transition can either occur spontaneously or be activated by temperature or by addition of adsorbates [10].

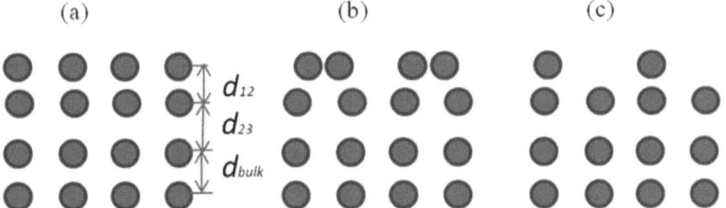

Figure 2.2. Schematic side view of characteristic rearrangements of surface atoms (a) relaxed surface ($d_{12} < d_{bulk}$), (b) Surface reconstruction (c) missing row type of reconstruction.

2.2.3 Adsorbate–induced reconstruction

At the lowest level of surface modification, adsorbates often influence the subtle changes in atomic layer spacings that occur close to the surface of a clean substrate observed by diffraction techniques. Certain surfaces (bare (110) Cu, Pd and Ni) are stable in the clean state while other surface reconstruct in the presence of adsorbed atoms. The (110) surfaces of transition metals Ag, Pt and Au transforms to (2×1) pattern of the missing row type of reconstruction (figure 2.2 (c)) when oxygen is adsorbed onto the surface [11–13]. The thermodynamic driving force for this kind of reconstruction is the formation of strong adsorbate–substrate bonds that are comparable to or stronger than the bonds between substrate atoms in the clean substrate surface.

2.3 Adsorption

The process of adsorption occurs when a gas or liquid solute accumulates on the surface of a solid forming a molecular or atomic film [14,15]. A qualitative distinction can be made between chemisorption and physisorption in terms of their relative binding strengths and mechanisms [11,16]. In the case of physisorption, the adsorbate–substrate interaction is weak due to van der Waals forces and the binding energies are within the ranges of 10 – 100 meV. The overlap of the wave functions of the molecule and the substrate is rather small, thus, the perturbations of the structural environment near the adsorption site is negligible [5,17]. When the adsorbate–substrate interactions forms strong chemical bonds that are either covalent or ionic with binding energies of about 1 – 10 eV, the process is often denoted chemisorption [11]. In the later process, the strong bonds alters the adsorbate chemical state and alternatively changes the structure of the substrate either by relaxation or reconstruction of the few top atomic layers of the substrate. This process is the result of the long range order observed on most single crystal surfaces with adsorbates which possess two–dimensional phase characterised by its own electronic, chemical, and mechanical properties [17,18].

2.4 Kinetics of adsorption

Consider the kinetic approach for the case of a uniform solid surface exposed to a monatomic gas of adsorbates at temperature T and pressure P. From the kinetic theory of gases P = nk_BT, the flux F of the gas molecules or atoms impinging on the surface per area and time is given by [19]

$$F = \frac{P}{\sqrt{2\pi m k_B T}} \quad (2.1)$$

where P is the partial pressure of the adsorption gas in torrs, m is the mass of the gas molecule (atoms) in Kg, k_B is Boltzmann's constant, and T is the temperature in Kelvin. The ratio of the adsorption rate to the incident rate is defined as the sticking coefficient. The flux is a measure of how fast material is deposited on the surface. It does not enter any equation concerning processes connected to self–organization, like the definition of diffusion barriers or the strain field. Thus, the flux rate influences the kinetics of growth only indirectly by determining the density of adsorbates on the surface and thereby the mean free path of adatoms, the nucleation rate of islands and by determining the time scale of free diffusion between two deposition events.

2.5 The $\sqrt{3}$ structures on (111) FCC metal surfaces

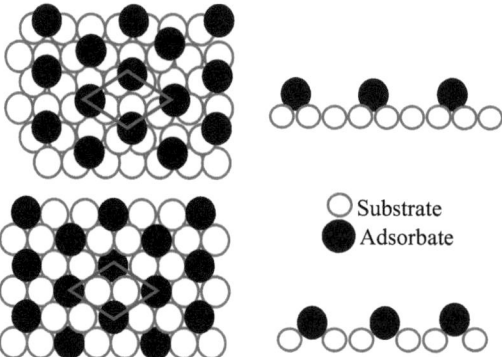

Figure 2.3. Schematic diagrams illustrating two possible surface structures that have a $(\sqrt{3} \times \sqrt{3})R30°$ structure, the $(\sqrt{3} \times \sqrt{3})R30°$ unit cell is outlined: (a) the adatoms sit on the hollow site of the FCC (111) plane, and (b) the adatoms are sitting substitutionally within the (111) surface layer showing rippling due to the outward shifting of the adsorbate atoms [20].

Even though real crystalline surfaces always contain point and/or line defects, the model of a periodic two–dimensional surface is convenient and adequate for the description of well–prepared samples with large well–ordered areas and low defect density. The coverage of foreign species adsorbed on surfaces of atomically clean substrates characterizes the surface concentration of the adsorbed material and is expressed in monolayer (ML) units. The adsorption of various metallic adsorbates on (111) surfaces of metals results in the formation of $(\sqrt{3} \times \sqrt{3})R30°$ structures [5, 20–23] at an adsorbate coverage of ~1/3 ML [14]. Depending on the adsorbate position on the substrate, the system can become a surface alloy i.e, substitutional $(\sqrt{3} \times \sqrt{3})R30°$ phase (figure 2.3 (a)) or alternatively adatom–type $(\sqrt{3} \times \sqrt{3})R30°$ phase (figure 2.3 (b)). In the prior phase, adsorbate atoms occupy substitutional sites formed by displacement of the substrate atoms from the first atomic layer of the substrate. Examples of substitutional structures include the adsorption of Sn, Sb and In on Cu(111), Ag(111) after annealing above 300°C [14, 20–23].

2.6 Dynamics of adatoms diffusion

Defects plays a prominent role in many properties of materials, through their characteristic production of local distortion and mobility [3,24,25]. The Terrace Ledge Kink (TLK) model (figure 2.4), which is also referred to as the Terrace Step Kink (TSK) model, describes the thermodynamics of crystal surface formation and transformation, as well as the energetics of surface defect formation [25,26]. Upon adsorption on the surface, adatoms has (positive) adsorption energy E_a, relative to zero in the vapour. (This energy is at times referred to the desorption energy).

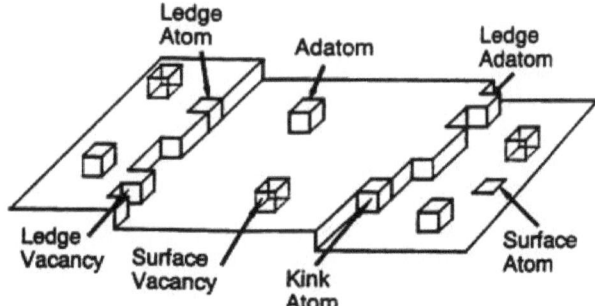

Figure 2.4. Schematic illustration of typical surface sites and defects on solid surfaces [25].

The desorption rate of adatom is roughly given by

$$r_d = v_a e^{-\left(\frac{E_a}{kT}\right)} \qquad (2.2)$$

where v_a is the characteristic atomic vibration frequency and is expected to vary relatively slowly (not exponentially) with temperature T, and k is Boltzmann constant. In this case, the desorption energy is assumed to come from lattice vibrations only. The adatom can diffuse over the surface, with energy E_d (migration barrier energy) and the corresponding frequency v_d (order of 10^{14} s^{-1}). Since $E_d \ll E_a$, surface diffusion is far more likely than desorption. The probability that during one second the adatom will have enough thermal energy to pass over the barrier is

$$P_j = v_d e^{-\left(\frac{E_d}{kT}\right)} \qquad (2.3)$$

In unit time the adatom makes v_d attempts to pass the barrier, with a probability of $e^{-\left(\frac{E_d}{kT}\right)}$ of surmounting the barrier on each try. The adatom diffusion coefficient (jumping a distance l) is then approximately

$$D = v_d l^2 e^{-\left(\frac{E_a}{kT}\right)} \qquad (2.4)$$

The diffusion of atoms on a lattice occurs by the random motion of atoms from lattice site to lattice site. Thus, equation (2.4) is in fact the mean square displacement of the random walker per unit time, or the tracer diffusion coefficient. It is convenient to express surface areas in terms of substrate unit cells. Then D becomes the number of unit cells visited by the adatom per unit time. The adatom lifetime before desorption is given by:

$$\tau_a = v_a^{-1} e^{-\left(\frac{E_a}{kT}\right)} \qquad (2.5)$$

Then the characteristic length, within which the adatom can move, is

$$L = \sqrt{D\tau_a} \qquad (2.6)$$

The presence of steps at the substrate surface can suppress the homogeneous nucleation of adatoms on terraces [27]. The growth mechanism will be in favour of heterogeneous step nucleation which in turn permits smooth two–dimensional growth. Binding energies for adatoms at step sites are in general larger than on terrace sites due to the increased coordination.

2.7 Interaction at the surface

2.7.1 Influence of adsorbates on surfaces

When atoms or molecules are adsorbed on the surface of a solid, the electronic structures of both the atom and the solid are perturbed. The electronic states on metallic surfaces are described by wave functions with periodic properties in directions parallel to the surface, and because of this, it is clear that the perturbation associated with the adsorption of an atom need not be very well localized. If this is so, then the adsorption of one atom will influence the adsorption of another over distances for which the direct interaction of the two atoms is negligible.

The electrons of the surface states on the (111) surfaces of noble metals (Au, Ag, and Cu) form a quasi two–dimensional free electron gas (2DEG) which is confined to the first few atomic layers at the crystal surface [28–30]. They are scattered by the potential associated with surface defects, e.g. impurity atoms, adatoms, or step edges, leading to quantum–interference patterns in the local density of states around these defects. Surface interactions such as the direct chemical interaction and the indirect electrostatic (dipole–dipole), elastic (deformation of substrate lattice) and electronic interactions (Friedel oscillations in the density of bulk conduction electrons), are present at a few atomic distances [28,31]. Friedel–oscillations at the Fermi energy (E_F) have periodicity of half the wavelength $\lambda = 2\pi k_F$ of screening electrons [32,33].

The Friedel type interaction between two adatoms on a Cu(111) surface is given by,

$$\Delta E_{int} = \frac{\cos(2k_F R + \varphi)}{R^n} \qquad (2.6)$$

where R is the distance between the adsorbed atoms, k_F the Fermi wavenumber, n a constant, and φ is the azimuthal angle between the wave vector projection on the xy plane and the surface normal [31]. The lateral range of the force, given by n, depends on the states providing the screening of the adatom. The theoretical result of current interest is that the interaction is particularly long range when the interaction is mediated by a partially filled surface band of electrons [31]. In crystalline surfaces of noble metals with broken structural inversion symmetry, the Fermi energy:

$$E_F = \frac{\hbar^2 k_F^2}{2m^*} \qquad (2.7)$$

where $\hbar$ is Planck's constant divided by 2π, k_F is the wave vector of the Fermi energy of an electron and m^* is the electron effective mass. The subtle interplay between orbital and spin moments through the spin orbit (SO) interaction induces a spin splitting of the electronic states in these systems. The resulting effect is called the Rashba–Bychkov effect [30] and the associated energy E_F or the strength of the spin orbit induced spin–splitting is called the Rashba energy [34–36]. Several studies [29,36–38] have shown that the splitting of the energy bands in the asymmetric surfaces can be enhanced by adsorption of various metals (Pb, Bi, and Sb) and various gases (O and Xe) on noble metals and these system have applications in the field of spintronics.

2.8 Types of growth

Augmented interest in studies of thin metal film growth is partly because of the possibility of growing thin films with novel physical and chemical properties and partly because of the desire to understand the fundamental processes underlying nucleation and growth. The growth behavior usually depends on the thermodynamics of the new solid phase formation and the growth kinetics [5,7,39]. Whereas thermodynamics tends to drive the system into its minimum free energy configuration, the trend towards thermodynamic equilibrium is often hindered by kinetic limitations. When metal atoms evaporate on a metal crystal, a supersaturated gas phase of metal atoms exists above the metal surface. The higher this supersaturation is, the more the kinetics will dominate the nucleation and growth. In general, on atomically flat surfaces, three principal modes of growth are established (figure 2.5).

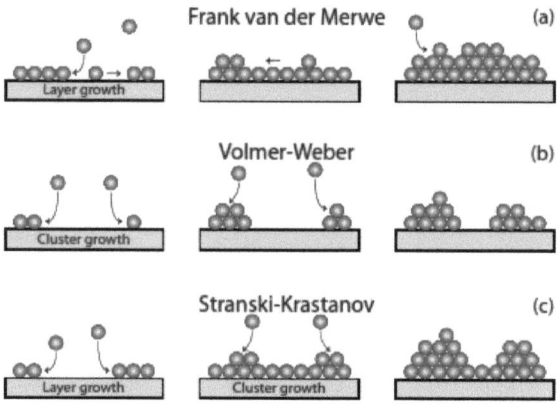

Figure 2.5. Schematic illustration of different growth modes.

2.8.1 Frank–van der Merwe (FM) or layer–by layer

The interatomic interactions between substrate and film materials are stronger and more attractive than those between the different atomic species within the film material. The FM growth requires that

$$\gamma_i + \gamma_f \leq \gamma_s, \qquad (2.8)$$

where γ_i, γ_f, and γ_s are the interface free energy, film free energy and surface free energy respectively.

2.8.2 Volmer–Weber (VW) or 3D island

This growth mode occurs when the atoms or molecules of the growing film are more strongly bonded to each other than to the substrate unlike in the FM growth. This type of growth is expected if

$$\gamma_i + \gamma_f > \gamma_s \qquad (2.9)$$

2.8.3 Stranski–Krastanow (SK) or 3D island–on wetting–layer growth

The SK growth mode illustrates the intermediate case between layer by layer and island growth. Strained monolayer (or several monolayers) of growth occurs first, with additional growth occurring in the form of islands nucleating on the growing film. As a result, there is a transition from two–to three–dimensional growth. Only when $\gamma_i + \gamma_f \sim \gamma_s$ is satisfied can this type of growth occur. In the SK growth, the lattice mismatch between the substrate and the absorbate induces a stress which is relieved after a number of layers by the formation of strain islands.

2.9 Introduction to Segregation

The process of segregation is defined as the diffusion of the segregant from the bulk to the surface. Its parameters are regarded as some energetics and diffusion factors that constitute surface segregation phenomenon. These are the segregation, activation and the interaction energies as well as the pre–exponential factor. By measuring these solute enrichments on the surface as a function of temperature or time, their segregation parameters such as the pre–exponential factor D_0 and the activation energy E could be extracted [40–46]. Consequently the data making up the equilibrium segregation profile can be utilized to determine other segregation parameters, namely, the interaction coefficient between the atoms *i* and *j*, Ω_{ij}, and segregation energy ΔG_i. At the initial stage, the segregation energy is responsible for driving the segregant atoms from the first bulk layer to the surface.

2.10 The modified Darken model

In the modified Darken model [47–49] it is assumed that the driving force in the segregating system is the gradient of the chemical potential instead of the concentration gradient as is assumed in the Fick description. The segregation process of a binary system is described by a set of coupled rate equations for a surface S in contact with the bulk consisting of N layers. In the original model [48] the net flux of species $i(J_i)$ through a plane at x = b is given by:

$$J_i = -M_i X_i^B \left(\frac{\partial \mu_i}{\partial x}\right)_{x=b} \quad (2.10)$$

where M_i is the mobility of the species *i* and X_i^B the supply concentration in between two layers (within the plane). This supply concentration from within the planes has no physical meaning and the first modification to the Darken model categorically associates the supply concentration to a specific layer as:

$$J_i^{(j+1,j)} = M_i X_i^{(j+1)} \frac{\Delta \mu_i^{(j+1,j)}}{d} \quad (2.11)$$

Equation 2.11 then indicates the flux of atoms from the $(j+1)$–th layer to the j–th layer with the supply concentration $X_i^{(j+1)}$ and the difference in the chemical potential between the layers $\mu_i^{(j+1,j)}$. The segregation system of surface ϕ and bulk B is therefore described by:

$$\frac{\partial X_i^\phi}{\partial t} = \left[\frac{M_i^{B_1 \to \phi} X_i^{B_1}}{d^2}\right] \Delta\mu_i^{(B_1,\phi)} \qquad (2.12)$$

And for the j–th layer,

$$\frac{\partial X_i^{(j)}}{\partial t} = \left[\frac{M_i^{j+1 \to j} X_i^{(j+1)}}{d^2} \Delta\mu_i^{(j+1,j)} - \frac{M_i^{j \to j-1} X_i^{(j)}}{d^2} \Delta\mu_i^{(j,j-1)}\right] \qquad (2.12)$$

for $i = 1, 2, \ldots, m-1$ and $j = \phi, B_{1,\ldots N}$. Now there are $(m-1)(N+1)$ rate equations for the $N+1$ layers.

2.11 References

[1] G.S Rohrer, *Structure and Bonding in Crystalline Materials*, Cambridge University Press (1998)

[2] D. Myers, *Surfaces interfaces and colloids*, Wiley Online Library (1999)

[3] D. P. Woodruff, *Surface Alloys and Alloys Surfaces*, Elsevier (2002)

[4] K. Binder, M. Bowker, J. E. Inglesfield, and P.J. Rous, *Cohesion and Structure of Surfaces*, Elsevier (1995)

[5] J. Venables, *Introduction to surface and thin film processes*, Cambridge University press (2000)

[6] D. M. Mattox, *Handbook of physical vapor deposition (PVD) processing*, Noyes Publications (1998)

[7] H. Ibach, *Physics of surfaces and interfaces*, Springer (2006)

[8] C. Kittel, *Introduction to Solid State Physics*, Wiley & Sons (1976)

[9] H.Y. Xiao, X.T. Zu, X.He, and F. Gao, Chem. Phys. 325 (2006) 519–524

[11] A. Nilsson, L. G. M. Pettersson, and J. K. Nørskov, *Chemical Bonding at Surfaces and Interfaces*, Elsevier (2008)

[12] D. P. Woodruff, J. Phys: Condens. Matter 6 (1994) 1869

[13] J. K. Burdett, P. T. Czech, and T, F. Fassler, Inorg. Chem. 31 (1992)

[14] K. Oura, V.G. Lifshits, A.A. Saranin, A.V. Zotov, and M. Katayama, *Surface Science: An Introduction*, Springer (2003)

[15] D. Woodruff, *The Chemical Physics of Solid Surfaces*, 10, Elsevier (2002)

[16] A. Gabor, *Introduction to Surface Chemistry and Catalysis*, John Wiley and sons (1994)

[17] A. Groß, *Theoretical Surface Science: A Microscopic Perspective*, Springer (2007)

[18] H. Ibach, *Physics of surfaces and interfaces*, Springer (2006)

[19] J. I. Gersten, and F. W. Smith, *The physics and chemistry of materials*, John Wiley and sons Inc (2001)

[20] B. Aufray, H. Giordano, and D. N. Seidman, Surf. Sci. 447 (2000) 180

[21] E. A. Soares, C. Bittencourt, V. B. Nascimento, V. E. de Carvalho, C. M. C. de Castilho, C. F. McConville, A .V. de Carvalho, and D. P. Woodruff, Phys. Rev. B 61 (2000) 13983

[22] S. A. de Vries, W. J. Huisman, P. Goedtkindt, M. J. Zwanenburg, S. L. Bennett, I. K. Robinson, and E. Vlieg, Surf. Sci. 414 (1998) 159

[23] T. C. Q. Noakes, D. A. Hutt, C. F. McConville, and D. P. Woodruff, Surf. Sci. 372 (1997) 117

[24] W. N. Unertl, *Physical structure*, Elsevier (1996)

[25] L. Hans, *Solid Surfaces Interfaces and Thin Films*, Springer 5th ed (2010)

[26] G. Dhanaraj, K. Byrappa, V. Prasad, and M. Dudley, *Springer Handbook of Crystal Growth*, springer (2010)

[27] C. Ratsch, J. Garcia, and R. E. Caflisch, Appl. Phys. Lett. 87 (2005) 141901

[28] N. Knorr, PhD thesis, University of Kaiserslautern (2002)

[29] C. R. Ast, J. Henk, A. Ernst, L. Moreschini, M. C. Falub, D. Pacile´, P. Bruno, K. Kern, and M. Grioni, Phys. Rev. Lett. 98 (2007) 186807

[30] Y. A. Bychkov, and E. I. Rashba, J. Phys. C: Solid State Phys. 17, (1984) 6039

[31] E. Wahlström, I. Ekvall, H. Olin, and L. Wallden, Appl. Phys. A 66 (1998) S1107–S1110

[32] P. A. Dowben, Mater. Res. Soc. Symp. Proc. 887 (2006)

[33] P. T. Sprunger, L. Petersen, E. W. Plummer, E. Lægsgaard, and F. Besenbacher, Science 275 (1997) 1764

[34] R. Winkler, and E. I. Rashba, Physica E 22 (2004) 450–454

[35] C. R. Ast, D. Pacilé, L. Moreschini, M. C. Falub, M. Papagno, K. Kern, and M. Grioni, Phys. Rev B 77 (2008) 081407

[36] L. Moreschini, A. Bendounan, H. Bentmann, M. Assig, K. Kern, F. Reinert, J. Henk, C. R. Ast, and M. Grioni, Phys. Rev. B 80 (2009) 035438-1 -035438-6

[37] L. Moreschini, A. Bendounan, I. Gierz, C. R. Ast, H. Mirhosseini, H. Höchst, K. Kern, J. Henk, A. Ernst, S. Ostanin, F. Reinert, and M. Grioni, Phys. Rev. B 79 (2009) 075424

[38] L. Moreschini, A. Bendounan, C.R. Ast, F. Reinert, and M. Grioni, Phys. Rev. B 77 (2008) 115407

[39] J. H. van der Merwe, R. Vanselow, and R. Howe, *Chemistry and Physics of Solid Surfaces*, Springer, Berlin (1984) 365

[40] J. du Plessis, and G. N. van Wyk, J. Phys. Chem. Solids 50 (1988) 247

[41] J. du Plessis, and G. N. van Wyk, J. Phys. Chem. Solids 50 (1988) 251

[42] J. du Plessis, and G. N. van Wyk, E. Taglauer, Surf. Sci. 220 (1989) 381

[43] G. N. van Wyk, J. du Plessis, and E. Taglauer, Surf. Sci. 254 (1991) 73

[44] J. du Plessis, and P. E. Viljoen, Surf. Sci. 276 (1992) 7

[45] J. du Plessis, Surf. Sci. 287-288 (1993) 857.

[46] E. C. Viljoen, J. du Plessis, H.C. Swart, and G.N. Van Wyk, Surf. Sci. 342 (1995)

[47] J. du Plessis, *Surface Segregation, Diffusion and Defect Data*, 11, Sci–Tech Pub. 9 (1990) 125

[48] L. S. Darken, Trans. AIME 180 (1949) 430

[49] H. Viefhuas, and M. Rüsenberg, Surf. Sci. 159 (1985) 1

CHAPTER 3

SCANNING TUNNELING MICROSCOPY (STM)

3.1 Introduction

STM studies of metal surfaces are still fairly scarce as compared to STM studies on semiconductor surfaces [1]. This is partly due to the localized nature of the charge densities and, the sp bonding surface states which present large corrugations near E_F. Metal surfaces have, on the other hand, more shallow corrugations dominated by delocalized sp states. In order to image the individual atoms on metal surfaces, a higher lateral and vertical resolution of the scanning tunneling microscope is thus a necessary requirement. Today this has become possible on a routine basis with the best and most stable STMs. As a consequence, there is a rapid increase on STM studies on metal surfaces [2–4].

Relying too heavily and uncritically on STM micrographs alone may in some instances lead to misleading conclusions. Thus in several cases, most STM studies are supported by conventional surface–sensitive techniques capable of providing for instance, crystallographic, electronic, subsurface and specific chemical information.

3.2 Electron tunneling

The ability of the STM to yield three dimension (3D) micrographs of surfaces with resolution of individual atoms and its ability to manipulate the atomic landscapes of conductive surfaces lies on the quantum mechanical phenomenon known as electron tunneling [4,5]. The concept of electron tunneling has been extensively studied theoretically and experimentally long before the invention of scanning tunneling microscopy [6]. Tunneling is a microscopic phenomenon where a particle can penetrate and in most cases pass through a potential barrier assumed to be higher than the kinetic energy of the particle. Such a motion is not allowed by the laws of classical dynamics and is described in detail in introductory quantum mechanics text books [7–9]. Electron tunneling originates from the overlap of wave functions between the molecules at the tip and surface atoms of the substrate (figure 3.1).The tunneling process may be considered as a charge–transfer process between electronic states of the tip and the sample.

Quantum mechanically speaking, the wave function of an electron describes its motions in all region of space. For example, if we consider the one dimensional tunneling model, the wave function of the electrons is governed by the Schrödinger equation given by:

$$-\frac{\hbar^2}{2m}\frac{\partial^2}{\partial z^2}\psi(z) + U(z) = E\psi(z) \qquad (3.1)$$

where m is the particle mass, ℏ is Plancks constant divided by 2π, and ψ is the wave function. The solution for the electron wave function in the classically allowed region E > U(z) is a plane wave

$$\psi(z) = \psi(0)e^{\pm ikz}, \qquad (3.2)$$

where z is the width of the tunneling barrier, m is the mass of the electron, U is the potential across the gap (the applied bias), and E is the energy of the electrons in this case the width of the gap and:

$$k = \frac{\sqrt{2m(E-U)}}{\hbar}, \qquad (3.3)$$

is the wave vector. The electron is moving (in either a positive or negative direction) with a constant momentum $p_z = \hbar k = [2m(E - U)]^{1/2}$, or a constant velocity $v_z = p_z/m$, the same as classical. In the classically forbidden region E < U(z), equation (3.2) has the following solution:

$$\psi(z) = \psi(0)e^{-\beta z} \qquad (3.4)$$

$$\text{where} \quad \beta = \frac{\sqrt{2m(U-E)}}{\hbar} \qquad (3.5)$$

is the decay constant which describes a state of the electron decaying in the +z direction. The probability density of observing an electron near a point z is proportional to $|\psi(0)|^2 e^{-2\beta z}$ which has a nonzero value in the barrier region, thus a nonzero probability to penetrate a barrier. Another solution, $\psi(z) = \psi(0)e^{\beta z}$, describes an electron state decaying in the –z direction.

3.3 STM principle of operation

To explore in real space the atomic–scale realm of surfaces in STM, a conducting atomically sharp tip (typically W or Pt–Ir alloy) mounted on three mutually perpendicular piezoelectric transducers (a ceramic material that expands or contracts in response to a change in the applied voltage) is brought into close proximity (~0.1 nm) of a metallic or semiconducting surface under applied bias voltage (V_{bias}) as in figure 3.1(a).

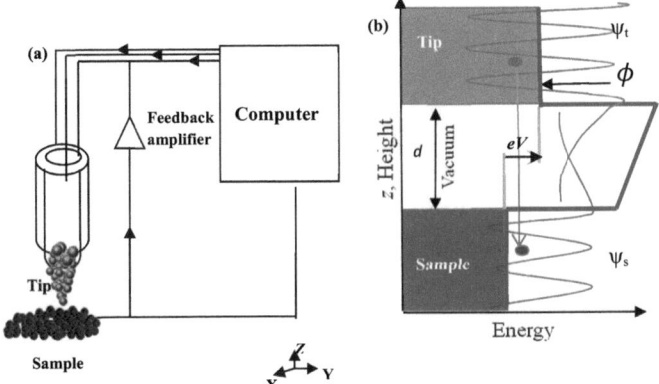

Figure 3.1. (a) Schematic representation of the STM. (b) Tunneling process between the tip and sample across a vacuum barrier of width d and height z. The electron wave functions ψ_t and ψ_s decay exponentially into vacuum.

The wavelike properties of electrons permit them to tunnel beyond the surface into regions of space that are forbidden to them classically but permitted quantum mechanically (figure 3.1(b)). This vacuum tunneling establishes a small tunnel current I_{tun} within the nano–ampere range [9,10]. When the tip–sample distance increases (figure 3.1), the probability of finding such tunneling electrons decreases exponentially (equation 3.7). The tunneling current of electrons is extremely sensitive to the distance between the last atom of the tip and the atoms of the underlying specimen. This gives rise to the local character of STM measurements, which makes it possible to visualize surface structures with sub–angstrom resolution and to detect various atomic–scale defects that are inaccessible by diffraction and spectroscopic techniques [11,12].

The measured tunneling current depends on the density of the electronic states, work functions of the tip – sample and the voltage biased. Because tunneling current is proportional to the local density of states (LDOS) near the Fermi level, the tip follows a contour of a constant density of states during scanning. The presence of adsorbates on the surface alters the LDOS in the surrounding area and the adsorbates are in general found to be imaged either as protrusions or depressions with respect to the bare surface. The output current differs when the probe tip is directly on top of an atom (smaller distance) as compared to when the probe tip is above spacing between atoms (larger distance).

3.4 Tunneling current

When there is no applied bias voltage ($V_{bias} = 0$), there is no net flow of electrons (figure 3.2 (a)) since the Fermi level E_F of both the tip and sample is equal, i.e., the gradient is zero. When $V_{bias} > 0$, (figure 3.2(b)) the Fermi level of the sample is raised by V_{bias}, the voltage difference between the tip and the sample creates an electric field, which causes electrons in the occupied state of the sample to quantum mechanically tunnel into the empty state of the tip. When the applied bias voltage $V_{bias} < 0$ (figure 3.2 c), electrons in the occupied state of the tip tunnel into unoccupied state of the sample. The electrons must satisfy Schrödinger's equation (3.1) as they tunnel across the tip–sample gap, whether it be in vacuum or ambient conditions.

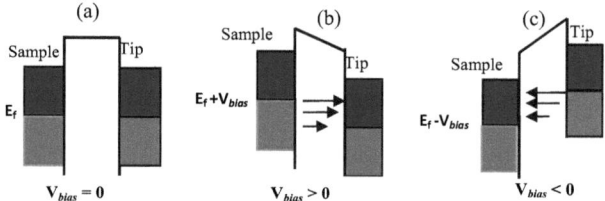

Figure 3.2. Schematic of a metal–insulator–metal tunneling junction at various bias voltages. The light blue area represents electron filled states and the brown area is the empty states.

The tunneling current is directly proportional to the number of surface electron states within an energy window (eV), where V is the voltage and E_F is the Fermi level, as shown below.

$$I \propto \sum_{E_F-eV}^{E_F} |\Psi(0)|^2 e^{-2\beta z} \qquad (3.6)$$

From equation 3.6 the probability of the electrons successfully crossing the gap between the tip and the sample (tunneling current) I is proportional to the tip–sample distance,

$$I \sim e^{-2kz} \qquad (3.7)$$

For most tip materials k ~1, which means for STM, the tunneling current increases (decreases) by about an order of magnitude for every decrease (increase) of the tip–sample distance by 1 Å respectively. This exponential dependence on distance allows for precise vertical measurements. Under small bias ($V < \phi$), the tunneling current can be expressed as

$$I = V\rho(E_F)e^{-1.0252\sqrt{\phi}z} \qquad (3.8)$$

where $\rho(E_F)$ is the local density of state (LDOS) around Fermi energy level, based on the assumption that a small bias voltage will not alter the LDOS near the Fermi level, the work function ϕ is in units of eV and the tip–sample separation z is in units of Å. The validity of this assumption is justified under usual STM conditions (weak coupling). In typical measurements of clean metal in UHV, where a bias voltage of around 100 mV or less is usually applied, which is much less than the typical work function of metals (4 – 5 eV). Experimentally, the apparent barrier height (ABH) can be measured by recording the tunneling current as a function of the tip–sample separation based on the following equation

$$\phi \approx 0.95 \left(\frac{\partial \ln I}{\partial z}\right)^2 \qquad (3.9)$$

As mentioned before, all these discussions are one dimensional. Considering that the lateral dimension of the tip apex is usually much larger than the decay length of the tunneling current (a few Å), this one-dimensional picture is rather accurate. The measured ABH can be used as an indicator of surface cleanliness, since a slightly contaminated metal surface will have a smaller work function than clean metal. There will usually be a charge transfer from metal to the adsorbed molecules lowering the work function.

3.5 Bardeen theory of tunneling

In contrast to the rather simple principle of electron tunneling in STM illustrated above, a complete theoretical description is comparably complex. With a few simplifications, a widely used basic theory for STM can be constructed describing elastic tunneling processes called the Bardeen approach to tunneling [6]. Bardeen's tunneling theory was published in 1961 and applied to the scanning tunneling microscope in 1983 by Tersoff and Hamann [13]. The approach considers the tunnel junction as two subsystems with wavefunctions χ and ψ.

The approach showed that the tunneling matrix element M is determined by a surface integral on a separation surface between the tip and the specimen, i.e.,

$$M = \frac{\hbar}{2m} \int_{z=z_0} \left(\chi^* \frac{\partial \psi}{\partial z} - \psi \frac{\partial \chi^*}{\partial z}\right) dS \qquad (3.10)$$

where ψ and χ are the wavefunctions of the two electrodes (tip and sample) [14]. The rate of electron transfer is determined by the Fermi golden rule. The probability w of an electron in the state ψ at energy level Eψ tunneling to a state χ of energy level E obeys the following equation:

$$w = \frac{2\pi}{\hbar} |M|^2 \delta(E_\psi - E_\chi) \qquad (3.11)$$

where E_ψ and E_χ is the energy of the states ψ and χ, respectively. The delta function implies electrons can only tunnel to and from states with the same energy. Tunneling without losing energy is known as elastic tunneling. The total tunneling current is thus the sum over all possible states. Including a bias voltage the current is

$$I = \frac{4\pi e}{\hbar} \int_{-\infty}^{\infty} [f(E_F - eV + \varepsilon) - f(E_F + \varepsilon)] \rho_1(E_F - eV + \varepsilon) \rho_2(E_F + \varepsilon) |M|^2 d\varepsilon \qquad (3.12)$$

where f(E) is the Fermi function and the density of states (DOS) are given by ρn. The Fermi distribution function act as a step function giving

$$I = \frac{4\pi e}{\hbar} \int_0^{eV} \rho_1(E_F - eV + \varepsilon) \rho_2(E_F + \varepsilon) |M|^2 d\varepsilon \qquad (3.13)$$

Assuming $|M|^2$ does not change much over the integration interval, it may be considered a constant and thus equation 3.13 becomes

$$I \propto \int_0^{eV} \rho_1(E_F - eV + \varepsilon) \rho_2(E_F + \varepsilon) d\varepsilon \qquad (3.14)$$

The tunnel current is the integrated product of the DOS of both electrodes. If both electrodes are normal metals, i.e. their DOS are flat, we see that the integral yields I α V. In most cases the interest is in the DOS of just one electrode, not the product of both. If one electrode has a constant DOS then the derivative of the tunneling current at small voltages is given by

$$\frac{dI}{dV} \propto \rho_1(E_F - eV) \qquad (3.15)$$

and this derivative conductance equation becomes the function of the metal DOS[15,16]. On adsorbate–covered metal surfaces, the measured STM contours become strongly bias–dependent. The applied bias voltage can be chosen in the millivolt regime for metals due to the lack of a bulk energy gap at the Fermi level. For a given tunnel current, a small bias is preferable for obtaining high spatial resolution, as a consequence of small tip–surface separation.

3.6 Modes of operation

3.6.1 Constant current imaging mode

The most common imaging mode is the constant current imaging mode (figure 3.3). In the later, as the tip is raster scanned across the sample surface, a feedback control loop is used to compare the measured tunneling current (I_{tun}) to a preset constant value (I_0), typically 0.5 – 5 nA.

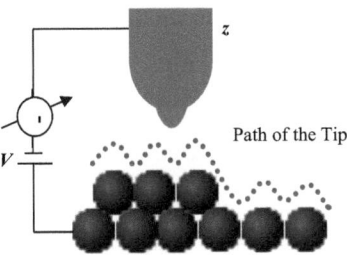

Figure 3.3. Schematic representation of the constant current mode.

If the measured current is less than the setpoint, the loop responds by moving the tip closer to the sample or else the tip retracts in case the measured current is greater than the setpoint value. The feedback signal, proportional to the difference between I_t and I_0, provides a correction voltage to the z transducer and thus causes the distance z between the tip and the surface to change when an atom is traversed. A topographic image of the sample surface with uniform electronic properties is generated by the computer recording the change in tip position z as a function of lateral position (x,y) [5,9].

The constant current topographs contain both geometric and electronic structure information. Generally, metals have the highest density of states where atoms are located, thus protrusions in STM micrographs corresponds to the location of atoms [17]. The constant current mode was the first mode historically developed for STM and its main advantage is the observation of surfaces which are not atomically flat [18].

3.6.2 Constant height imaging mode

Alternatively, in the constant height mode (figure 3.4), the tip scans the surface of the sample at nearly constant height and constant voltage while the tunneling current varies [18,19].

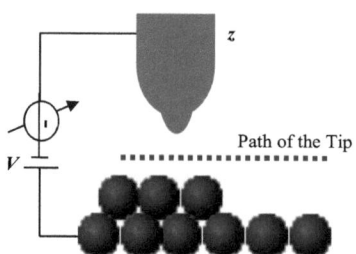

Figure 3.4. Constant height mode illustration.

In this mode the feedback mechanism is switched off. Consequently, as the current varies with the electron density of states, monitoring of the probe current as a function of x–y displacement yields a topographical representation of the surface morphology. In this mode, much faster imaging of atomically flat surfaces can be achieved since the feedback loop and the piezoelectric driver does not have to respond to the surface features passing under the tip [18].

Typically, in commercial scanning probe microscopes, each scan line consists of 512 points and scanning proceeds with line scanning frequencies in the 1–60 Hz range. For scanning flat areas of the order of 40 nm × 40 nm or smaller, where atomic and molecular-scale images are collected, scan line frequencies in the 8–60 Hz range are commonly used. Some of the distinct advantages of fast imaging include the observation of dynamic processes in real time and at reduced data collection time and minimal image distortion [18,19].

3.7 Scanning Tunneling Spectroscopy (STS)

STS is an extension of the tunneling junction experiment [6,11,20,21], but with a richer information content. Actually, the inception of the STM was driven by an idea to perform tunneling spectroscopy locally on an area less than 100 Å in diameter [22]. A concrete suggestion of a scanning tunneling spectroscopy (STS) experiment was made by Selloni et al. [23] on graphite and the actual experiments were latter done by Feenstra and co–workers [24,25].

Several review articles and books [26–28] have been published with extensive information about the technique; however, what follows is a brief summary of the relevant points. The STS measurements are taken while the feedback loop of the STM is temporarily disabled. While the tip is at a constant height, the bias voltage is varied through a range (usually of a few volts), and the current generated by the bias voltage is recorded. This procedure generates a current–voltage (I–V) curve, from which a great deal of information can be gathered [29]. The I–V curve can be manipulated in order to generate the differential conductance curve $\frac{dI}{dV} vs V$, as well as the normalized differential conductance curve, $\frac{VdI}{IdV} vs V$ which both which provides information that is the convolution of the tip DOS and the sample DOS when the bias voltage range is lower than the work functions of both tip and sample. [26,30,31].

3.7.1 Current Imaging Tunneling Spectroscopy (CITS)

Current Imaging Tunneling Spectroscopy (CITS) is a method of taking data where topography and I–V are taken simultaneously [32]. A scan line for the topography is taken by raster scanning the tip across the sample. The tip is moved back along the scan line and full I–V curve at each pixel of a topographic micrograph are taken. When performing CITS, each I–V consists of a single voltage sweep and each voltage sweep has a number of data points. Feed–back circuit interrupted and bias voltage ramped V reset to its original value and progress along the raster scan. The topography I–V correlation provided by this method provides a powerful diagnostic tool, in particular, regions of different local electronic structure can be identified [33, 34]. By adding a modulation voltage to the bias voltage, a lock–in amplifier can be used to simultaneously measure the differential conductance, dI/dV. Assuming the density of states of the tip is flat the dI/dV is proportional to the LDOS as shown in Eq. 3.15. More generally, dI/dV is related to the convolved tip and sample local density of states (LDOS). Like the topography data, I–V data is affected by the same issues discussed below which influences STM measurements.

3.7.2 Constant Current Spectroscopy

While constant height spectroscopy can reveal a great deal of information about a sample, it is limited by the dynamic range of the STM's *I–V* converter. Constant current spectroscopy is useful for acquiring spectroscopic information over a voltage range that is not limited by the STM's *I–V* converter.

In this mode, the STM feedback loop remains active and maintains a constant current as the voltage is varied. In this mode, dI/dV and the tip height, $z(V)$ are simultaneously recorded [35]. Although dI/dV is no longer simply proportional to the LDOS, peaks in dI/dV (and corresponding steps in $z(V)$) indicate regions where the LDOS rapidly changes with voltage, such as tunneling resonances or band edges.

3.7.3 Constant Voltage Spectroscopy

This mode is useful for estimating the sample work function. The tunneling current is recorded as a function of the tip height z. At low voltage, and assuming constant electric field within the junction, one finds that:

$$\frac{d\ln I}{d\delta(z)} = \frac{2\sqrt{2m\phi_a}}{\hbar} \qquad (3.16)$$

where m is the electron mass and ϕ_a is an effective barrier height [1]. Thus, by plotting ln I versus z, we expect a straight line whose slope can be used to calculate ϕ_a. While it is difficult to directly relate ϕ_a to the sample work function ϕ_s [7], adsorbate–induced changes in ϕ_a generally correlate with changes in f ϕ_s measured using other techniques. A dI/dV image maps out the location of the density of electrons at a particular energy, i.e. it images the differential conductivity at certain energy over the surface of the sample. Most surfaces do not have a spatially varying density of states at a given energy and therefore a dI/dV map on such a surface would be uniform. However, certain conditions such as a surface state band edge or adsorbates on a surface can create a spatially varying density of states at a given energy.

3.8 STM Major Components

To achieve atomic resolution with the STM, major challenges lies in the following formidable experimental problems: isolation of the experiment from natural vibrations of ~ 100 Hz that are present in every laboratory [36], tip to sample position control with sub angstrom precision, tip sharpness and quality of the sample surface.

3.8.1 Vibration Isolation

Since tunneling current depends exponentially on the gap between the tip and the sample, thus as a rule of thumb, even minute vibrations, such as those caused by noise and people walking around the building can hinder the achievement of atomic resolution and the stability of the STM system. The typical corrugation amplitude for STM images is about 0.1 Å. Therefore, the disturbance from external vibration should be reduced to less than 0.01 Å [12].

To reduce the influence of external vibration when designing the instrument, Binnig and Röhrer employed a vibration isolation system which utilizes soft metal springs with resonance frequency smaller than the frequencies coming from the surrounding and permanent magnets [22,37]. The principle is based on eddy current damping mechanism. As the magnets (fixed on the STM stage) move relative to a conductive block of copper, eddy currents are induced in the metal [12,37]. A magnetic field caused by these currents acts oppositely on the magnets/STM stage and dampens its motion in all directions.

3.8.2 Electronic feedback control system

The key to scanning tunneling microscopy is the control of the position of the tip probe with respect to the sample surface. To address the complexity of control and scanning movement of the probe tip in three spatial dimensions defined in figure 3.1(a), the tip is mounted on a piezoelectric tube scanner that can be driven with high accuracy by electronic feedback control system [37]. Other functions of the electronic feedback control system includes ensuring the presence of a tunneling current between the tip and the sample surface, acquisitions of surface structure data and interfacing with the computer controller in order to store and analyze the acquired data.

3.8.3 The STM Tip

The STM tip geometry (sharpness), shape (mechanical rigidity), and chemical composition (cleanness) are the key features which determine lateral resolution and reproducibility of STM scans [38–41]. In order to resolve small features such as atoms at the surface of a sample, it is necessary to utilize a probe tip ideally comparable in size to the features at the surface [41].

The electrical characteristics of the tip as well as the environment where it will be used (ambient or vacuum) should be taken into consideration when selecting the tip material. Transition metals are superior candidates because their d–band electron orbitals are more pointed than the s–, p–, or f–band orbitals and thus the tunneling current is predominantly contributed by the d state [42,43]. Tungsten is very hard, has high electrical conductivity and it can be easily electro–chemically etched in KOH or NaOH solutions to a fine point [44,45]. In UHV conditions, tungsten (W) tends to fit the necessary requirements for a probe tip and is commonly used in STM measurements.

3.8.4 Images and Filtering

In STM images, the spatial variation of the tip–height or the strength of the tunneling current is represented by a colour–coded contrast. When measurements are taken in constant current mode, the contrast refers to the spatial variation of the z–height of the tip, and such an image is called the height image. For operation in the constant–height mode, the image contrast refers to the spatial variation of the strength of the tunneling current. The acquired images are dubbed current images and often exhibit a higher contrast of the morphological and atomic–scale details than the height images. However, the latter provide more correct topographic information. In this work, intensity scale images of both types are presented.

Brighter spots correspond generally to elevated surface regions in height images, and to places with stronger current interaction in current images.

During STM measurements, it is common to observe variations of the image details in atomic–scale patterns depending on the experimental conditions [46,47]. Image variation may occur spontaneously due the instability of the tip and the surface, even when the scanning is carried out without changing the experimental conditions. Alternatively, image variation may reflect a change in the tip–sample distance and tip–sample interaction. If a sample surface has structural features sharper than the tip apex, the imaging roles of the tip and surface are reversed so that the tip shape appears in the image [48–50]. In many cases, the as–received STM images are treated to give a better presentation. A plane fit adjustment is necessary when the sample surface under examination is not exactly perpendicular to the scanner z–axis. This and other procedures of image modification (e.g., low pass and high pass filtering, flattening, erasing scan lines) are typically included in the scanning probe image processor software package (SPIP). This filtering highlights the periodic features of the image but loses information about the non periodic features that might have resulted from local defects.

3.9 Atomic resolution on well known substrates

The analysis of surface features from tunneling images naturally brings into question the interpretation of the source of the symmetry and apparent heights displayed in STM images. Considering that tunneling depends not only the tip–sample distance but also the electronic properties of the tip and sample materials, it is worth noting that constant current STM images contain valuable geometric and electronic structure information of materials. This chapter explores the surfaces of well known materials such as HOPG and Si(111) studied at the atomic scale.

3.10 Highly Oriented Pyrolytic Graphite (HOPG)

Highly Oriented Pyrolytic Graphite (HOPG) is a unique form of graphite which has preferable orientation in the (0001) plane [37–39]. Its crystal structure is characterized by an arrangement of carbon atoms (planar sp^2 bonds whose C–C bond length is about 0.142 nm) in stacked parallel layers called graphene (figure 3.5).

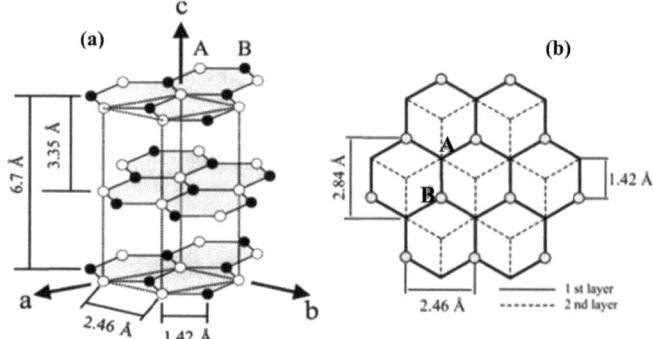

Figure 3.5. (a) Schematic representation side view of the structure of the bulk hexagonal graphite crystal (ABAB stacking), showing the two non–equivalent types of carbon atom sites: The in–layer nearest carbon–carbon distance is 0.142 nm and carbon layers are separated by 0.335 nm. (b) Top view of the two top most graphene layers [43].

These planes are held together by weak van der Waals force, thus the ease to cleave HOPG and basis for the writing capabilities of a pencil [40–42]. This is an easily renewable material with an ideal atomically flat surface and provides a background with only carbon in the elemental signature thus making results in a featureless background which is vital for scanning probe microscopy (SPM) measurements that require uniform, flat, and clean substrates for both calibration and/ or growth of nanostructures. The lattice of graphite consists of two equivalent interpenetrating triangular carbon sublattices and, each one contains a half of the carbon atoms (figure 3.5).

Each atom within a single plane has three nearest neighbours: the A–type carbon atoms lie directly above the carbon atoms of the layer beneath and they share their electron density, while the B–type carbon atoms are located above the centers of the carbon hexagons of the underlying layer [38–41]. The B–type carbon atoms show a higher electron density near the Fermi level compared to A–sites as evident from STM images. Since the STM senses the tunneling current from local density of states (LDOS), the differences in LDOS is relevant in STM. The carbon atoms in each graphene layer form a grid of correct hexagons with 0.142 nm distances between atoms and the layers are 0.35 nm apart [43].

3.10.1 Experimental

The HOPG samples were 1 cm^2 of ZYH grade from Veeco. Samples were prepared by the scotch–tape pilling method in air and immediately introduced to the UHV chamber without breaking vacuum. The STM observations were carried out with a commercial

Ultra–High Vacuum Variable Temperature STM (Omicron NanoTechnology GmbH) operated in the constant current imaging mode at a base pressure of 10^{-10} Torr at room temperature. The probe tips were prepared by electrochemical etching a 0.3 mm tungsten wire on NaOH solution. Further details on the STM tip fabrication are given in **Chapter 4.**

3.10.2 Results and Discussion

There has been many proposed mechanism in explaining the anomalies observed on HOPG surfaces [44–49], however, this section will only focus on the measured inter atomic distances between carbon atoms on HOPG as compared to the reported literature.

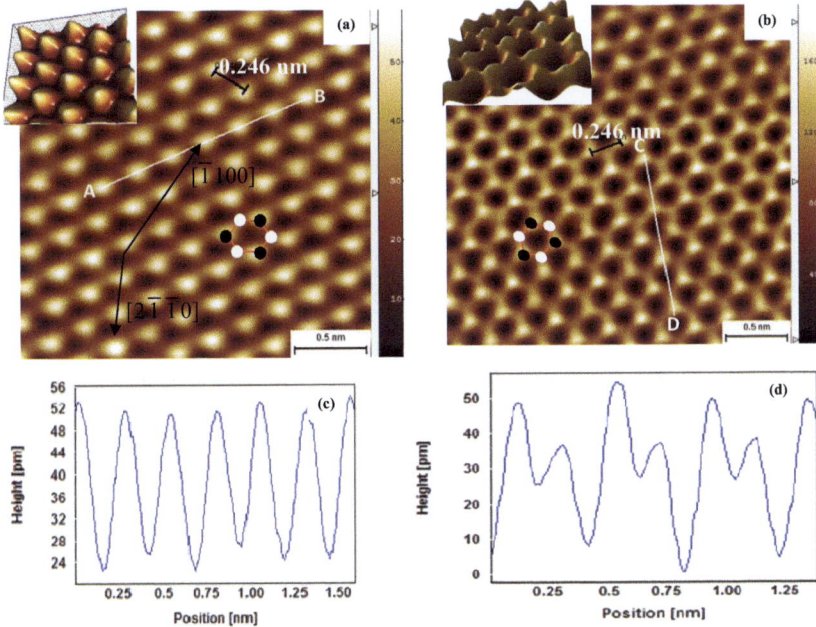

Figure 3.6. Topographic constant–height STM images of graphite (V_b –0.4 V, I_t 2 nA). One hexagonal surface unit cell with the two basis atoms A (white) and B (black) is superimposed in each image for clarity. (a) Typical STM image of HOPG showing the trigonal lattice structure (2.45 nm × 2.45 nm). (b) Ideal image (2.44 nm × 2.44 nm) showing honeycomb rings of the hexagonal structure, both A and B–type carbon atoms are imaged. The inserts on the top left show the 3D view of each image. (c–d) Illustrates the line profile in taken along the lines in (a) and (b) respectively.

Figure 3.6 shows atomic resolution of pristine clean surface of HOPG studied by ultra high vacuum VT–STM system. Both triangular and the honeycomb structures were obtained simultaneously after a few minutes of scanning. Figure 3.6 (a) depicts a regular triangular structure showing every second (B–type) carbon atom in each hexagon of graphite which is attributed to the non–equivalency of the graphite sites. The periodicity of the triangular structures is about 0.246 ± 0.01 nm, which is in agreement with the spacing between two neighboring B–site carbon atoms as represented in the line scan in figure 3.6 (c). Six nearest equidistance neighbours surround each apparent atom. In figure 3.6 (a) the carbon atoms appear as elliptical spots having their long axis along the [$\bar{1}$ 100] lattice direction. The image show a 2–fold symmetric pattern where every second carbon atom is missing. The corresponding line scan (figure 3.6 (c)) show clearly pronounced maxima with almost the same height corrugation. Figure 3.6 (b) depicts the regular graphene honeycomb structure presented in hexagonal rings with lattice distance of 0.142 ± 0.02 nm corresponding to the atomic distance between two adjacent AB carbon atoms in graphite. This is further illustrated by the corresponding cross section scans in figure 3.6 (d). The regular graphene honeycomb structure, displays complete graphite surface lattice (figure 3.6 (b)) with visible six carbon atoms on each hexagon (inset in figure 3.6 (b)). The centers of the hexagonal rings appear as dark holes. The distance between the centers of the holes is 0.246 ± 0.01 nm. The bright spots with high intensity correspond to the B–site carbon atoms while the A–site atoms appear as saddle points. The images were easily reproducible on different samples and with different W tips. The STM results obtained on HOPG are consistence with other experimental findings on graphite as in the schematic of figure 3.5 in the distance between the B type atoms is 0.246 nm.

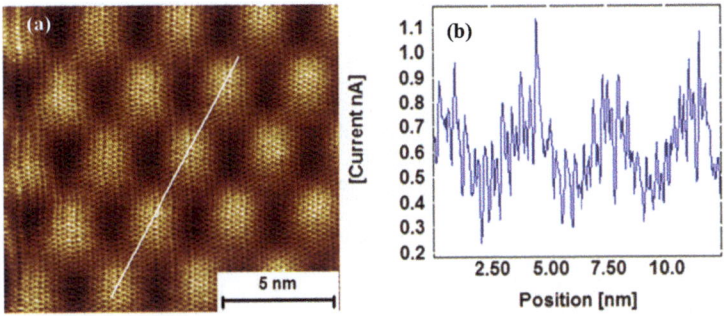

Figure 3.7. (a) STM image of Moiré patterns on HOPG. V_b : 0.3 V, I_t: 2.7 nA. (b) Line trace taken along the line in (a) showing the amplitude of the patterns

When there is a mis-orientation of graphene layers by an angle, a superlattice which leads to the formation of an interference pattern known as a Moiré pattern - with the same symmetry, but a super period is obtained (figure 3.7 (a)). The observed STM contrast of Moiré pattern comes from the strong influence of the layers on surface electronic structures near the Fermi level. Since the HOPG was not treated chemically or thermally in these studies, the top most graphele layer must have been rotated during cleaving. In these structures, both the sublattice symmetry and the linear band dispersion are preserved over the unit cell which is contrary to the AB (Bernal) stacking of bilayer graphene. The periodicity of the superlattice is about 3.45 ± 0.02 nm. The line trace in figure 3.7 (b) depicts the corrugations of the Moiré patterns.

3.11 Si(111) – 7×7 reconstruction

The surfaces of Si(111)–7×7 is one of the most complicated and fascinating object of investigation in both theoretical and experimental materials science [50–52]. The large unit cell (2.76 nm × 2.76 nm) of the reconstructed surface provides a platform for the testing of the unprecedented resolution of STM and is an ideal template for the growth of well-ordered nanostructures. The first real space atomic image of this surface was obtained by Binnig et al. in their landmark STM experiment [53], in which twelve bright spots corresponding to the topmost adatoms are revealed.

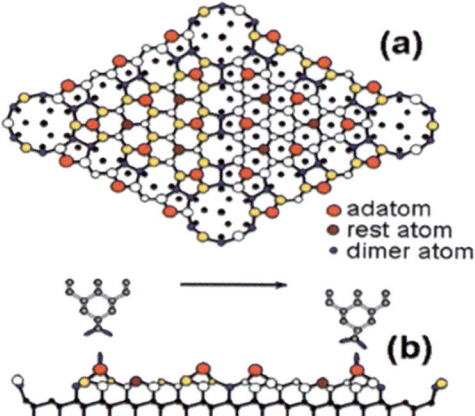

Figure 3.8. Schematic diagram for Si(111)–7×7 "DAS" model [7]. (a) Top view: atoms on (111) layers with the decreasing heights indicated by dots of decreasing sizes. (b) Side view: dangling bonds are located at the topmost of all adatoms, rest atoms and holes [55].

The accepted model (figure 3.8) in explaining the geometry of the Si(111)–7×7 surface reconstruction is the dimer–adatom–stacking fault (DAS) proposed by Takayanagi et al. [54]. The DAS model outlines the several features of the energetically stable unit cell of the (7×7) reconstruction (figure 3.8). These includes 12 adatoms per 7×7 unit cell, a hole in each corner, 3 dimers (two bounded atoms) on each side of the triangular halves, stacking fault in one of the halves, and 19 dangling bonds in each unit cell (reduced from 49 of the unreconstructed surface). The minimization of the dangling bonds is the cause of the reconstruction or rearrangement of surface atoms, which differs from the bulk. The periodicity of the reconstructed surface is seven times that of the bulk terminated surface structure, and thus the name (7×7). The side view shows that the stacking sequence in the right half of the unit is the same as in bulk Si(111) while the stacking sequence in the left half is faulted [56].

3.11.1 Experimental

The sample was an n–type silicon strip of 1 mm^2 and 0.5 mm thick, (111)–orientation, ~0.3 Ω cm conductivity, from Omicron NanoTechnology. The Si sample was outgassed by resistive heating at 630°C for 12 hours in UHV. After degassing, the sample was annealed by direct current heating to 850°C while keeping the chamber pressure below 1×10^{-10} Torr. Sample temperature was increased instantly to 1200°C (flash) for 20–60 sec, in order to clean the surface of contaminants (particularly carbon and oxygen). After the application of 3–4 flashes the sample is cooled at 1°C/s to 600°C while monitoring the pressure and then slowly decreasing the temperature to room temperature. This process yields pristine clean surfaces and nearly perfect (7×7) reconstruction confirmed by AES and LEED.

3.11.2 Results and Discussion

Figure 3.9 depicts a representative STM images of the (7×7) reconstructed surface imaged at room temperature. All images were measured in constant current mode with W tip. The intensity scales in the images correspond to a height difference. The white protrusions correspond to the outermost surface atoms (adatoms). The black lines outline the rhombic unit cell. It is known from literature that the two triangles building the rhombic unit cell are not identical. Below one triangle (FHUC), a stacking fault occurs in the layer below the 6 adatoms which appear bright. The other triangle (UHUC) in the unit cell has no stacking fault relative to the bulk structure. Each of the unit cells has dimensions of 2.76 ± 0.01 nm (figure 3.9). Figure 3.9 (a) displays a negative bias (–1.7 V) STM image of the spatial distribution of occupied surface electronic states (filled–state) of Si (111) 7×7 surface over an area of 19.1 nm × 15.2 nm. The image reveals simultaneously the 12 adatoms and 6 rest atoms in each rhombic unit cell. Further, in each corner of the cell a hole exists (corner hole). Each depression or corner hole has 6 of the protruding Si adatoms surrounding it. The faulted half of the 7×7 unit cell (FHUC) can be distinguished from the unfaulted half (UHUC) by their brighter appearance in the filled sate image.

The contrast difference is because there is more charge on the adatoms in the faulted half of the unit cell compared to those in the unfaulted half [52,55]. In both the FHUC and

UHUC, the corner adatoms (marked 1 and 6) and the centre adatom (marked 3 and 4) exhibit stronger corrugations while the rest atoms (marked 2 and 5) appears as saddle points. This is because the electronic states of adatoms are closer to Fermi level than those of the rest atoms

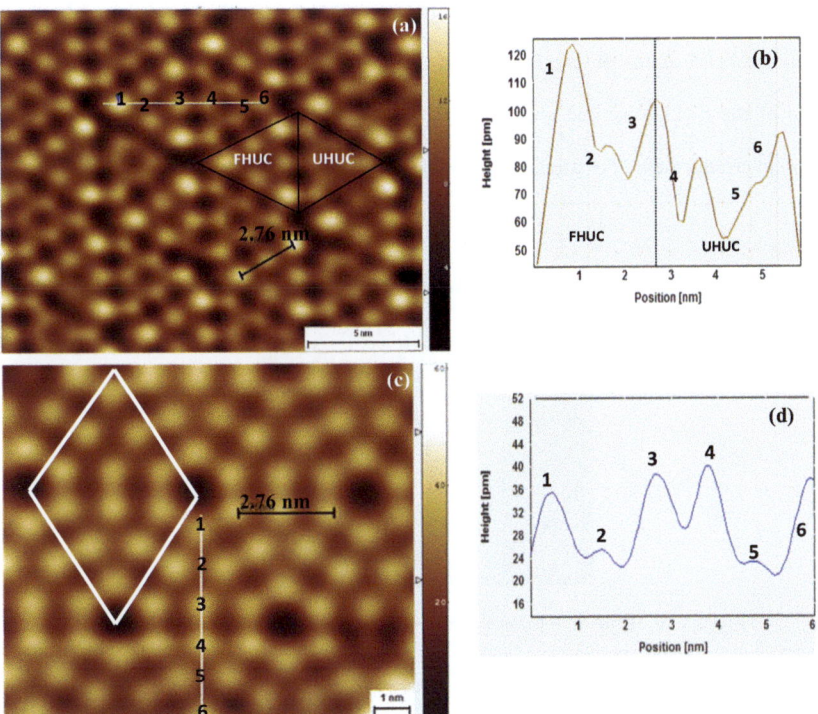

Figure 3.9. (a) High resolution STM image of occupied states of Si(111) – 7 × 7 surface revealing 12 adatoms and 6 rest atoms per (7 × 7) unit cell. The image was recorded by sample bias voltage of –1.7 V and tunneling current of 0.4 nA. (b) The line profile taken along the line in (a). Labels "1," "2," and "3" denote the corner adatom, the rest atom, the center adatom in the FHUC, and labels "4", "5", and"6"denote the center adatom, the rest atom, the corner adatom in the UHUC, respectively. (c) Unoccupied states STM image taken at a bias of 1.7 V and 0.4 nA (11.7 nm × 10.5 nm).

This effect is illustrated more clearly in the cross–sectional profile in figure 3.9 (b) taken along the image in figure 3.9 (a). The missing corner adatom defect seen at the surface (figure 3.9 (a)) has no influence on the position of its adjacent rest atom which is still imaged by STM. In the accompanying figure 3.9 (b), the high resolution un–occupied surface states STM image taken with a positive bias (1.7 V). The line profile in figure 3.9 (d) taken along the line in figure 3.9 (c) denotes the positions and asymmetry of both the adatoms and rest atoms on the FHUC and the UHUC. Figure 3.10 illustrates the LEED pattern of the Si(111) sample acquired after cycles of flash cleaning of the sample. The hexagonal arrangement of Si atoms is readily seen from the bright diffraction spots.

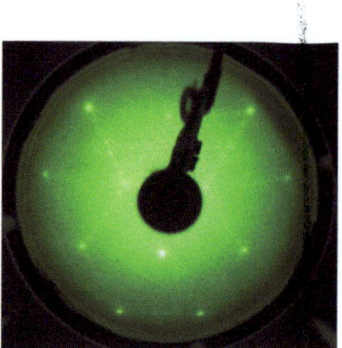

Figure 3.10. Diffraction pattern of Si(111) taken at 68 eV after flash cleaning.

3.12 References

[1] R. M. Feenstra, *Scanning Tunneling Microscopy and Related Methods*, Kluwer, Dordrecht (1990)

[2] D. Y. Kang, J. H. Lee, B. K. Oha, and J. W. Choi, Biosensors and Bioelectronics 24 (2009) 1431–1436

[3] T. T. Tsong, *Atom–probe Field Ion Microscopy*, Cambridge: Cambridge University Press (1990)

[4] M. Uiberacker, Th. Uphues, M. Schultze, A. J. Verhoef, V. Yakovlev, M. F. Kling, J. Rauschenberger, N. M. Kabachnik, H. Schröder, M. Lezius, K. L. Kompa, H.-G. Muller, M. J. J. Vrakking, S. Hendel, U. Kleineberg, U. Heinzmann, M. Drescher, and F. Krausz, Nature 446 (2007) 27-632

[5] G. Binnig, and H. Rohrer, IBM Journal of Research and Development 44, 1 (2000)

[6] J. Bardeen, Phys. Rev. Lett. 6, 57–59 (1960)

[7] E. Merzbacher, *Quantum Mechanics*, John Wiley & Sons, New York (1998)

[8] D. Griffiths, *Introduction to quantum mechanics*, Prentice Hall (1995)

[9] M. Razavy, *Quantum Theory of Tunneling*, World Scientific Printers (2003)

[10] R. Wiesendanger, *Scanning Probe Microscopy and Spectroscopy*, Cambridge University Press (1994)

[11] C. Julian Chen, *Introduction to Scanning Tunneling Microscopy*, 2nd Edition, Oxford University Press (2008)

[12] D. W. Pohl, IBM J. Res. Develop. 30, 4 (1986)

[13] J. Tersoff, and D. R. Hamann, Phys. Rev. Lett. 50, 25 (1983)

[14] G. Hormandinger, Phys. Rev. Lett. B, 49, 19 (1994)

[15] C. J. Chen, and T. J. Watson, J. Vac. Sci. Technol. A 62(1988)

[16] C. Wagner, R. Franke, and T. Fritz, Phys. Rev. B 75, 235432 (2007)

[17] D.P. Woodruff, *The chemical physics of Solid Surfaces*, Surface Alloys and Alloys Surfaces, Elsevier, 10 (2002)

[18] P. Hansma, J. Tersoff, J. Appl. Phys. 61, R1R (1986)

[19] C. Bai, *Scanning Tunneling Microscdopy and its Applications*, Springer Series in Surface Science 32 (1995)

[20] I. Giaever, Phys. Rev. Lett. 5 (1960) 147–148

[21] I. Giaever, Phys. Rev. Lett. (1960) 464–466

[22] G. Binnig, and H. Rohrer, Rev. Mod. Phys. 59, 615 (1987)

[23] A. Selloni, P. Carnevali, E. Tosatti, and C. D. Chen, Phys. Rev. B 31, (1985) 2602–2605

[24] R. M. Feenstra, Phys. Rev. B 44 (1991) 13791–13794

[25] R. M. Feenstra, A. J. Slavin, G. A. Held, M. A. Lutz, Phys. Rev. Lett. 66 (1991) 3257–3260

[26] R. M. Feenstra, J. A. Stroscio, and A. P. Fein, Surf. Sci. 181, 295 (1987)

[27] R. M. Tromp, J. Phys. Condens. Matter 1 10211 (1989)

[28] R. M. Feenstra, Surf. Sci. 299-300 (1994) 965

[29] R. M. Feenstra, W. A. Thompson, and A. P. Fein, Phys. Rev. Lett. 56, 608 (1986)

[30] D. Malterre, B. Kierren, Y. Fagot–Revurat, S. Pons, C. Didiot, H. Cercellier, A. Bendounan, New Journal of Physics 9 (2007) 391

[31] J. A. Stroscio and W. J. Kaiser (ed) *Scanning Tunneling Microscopy*, Boston, MA: Academic (1993)

[32] R. J. Hamers, R. M. Tromp, and J. E. Demuth, Phys. Rev. Lett. 56, 1972 (1986)

[33] T. Berghaus, A. Brodde, H. Neddermeyer and St. Tosch, J. Vac. Sci. Technol. A 6, 483 (1988)

[34] A. Basu, A.W. Brinkman, R. Schmidt, Z. Klusek, P. Kowalcyzk, and P. K. Datta, J. Eur. Ceram. Soc. 24, 1149 (2004)

[35] M. Ziegler, N. Néel, A. Sperl, J. Kröger, and R. Berndt, Phys. Rev. B 80, 125, 402 (2009)

[36] M. Bowker, P. R. Davies (Editor), *Scanning Tunneling Microscopy in Surface Science*, Willey (2009)

[37] Omicron NanoTechnology, VT STM User's Guide Manual, 4.0 (2007)

[38] J. Tersoff and D. R. Hamann, Phys. Rev. B 31, 805 (1985)

[39] E. Stoll, A. Baratoff, A. Selloni and P. Camevaffi, J. Phys. C 17 (1984) 167

[40] V. T. Binh and J. Marien, Surf. Sci. 202(1–2) (1988) L539–L549

[41] A. J. Melmed, J. Vac. Sci. Technol. B 9, 2, (1991)

[42] R. A. Deegan, Phys. Rev. 171 (1968) 659–664

[43] C. J. Chen, *Introduction to Scanning Tunneling Microscopy*. New York: Oxford University Press (1993)

[44] Omicron NanoTechnology, Tip Etching Kit User Manual, Version 1.1 (1998)

[45] A. J. Melmed, J. Vac. Sci. Technol. B 9, 601 (1990)

[46] R. M. Trompt, J; Phys. Condens. Mutt. (1989), I, 10211

[47] H. A. Mizes, S. Park, W. A. Harrison, Phys. Rev. B 36 (1987) 4491

[48] S. N. Magonov, H. J. Cantow, M. H. Whangbo, Surf: Sci. Lett. 318 (1994) L1175

[49] S. N. Magonov, A. Ya. Gorenberg, H. J. Cantow, Polyrn. Bull. 28 (1992), 577

[50] T. R. Albrecht, H. A. Mizes, J. Nogami, S. Park, C. F. Quate, Appl. Phys. Lett. 64 (1994) 1738

[51] I. Stich, M. C. Payne, R. D. King–Smith, J. S. Lin, L. J. Clarke, J, Phys. Rev. Lett. 68 (1992) 1351.

[52] K. D. Brommer, M. Needels, B. E. Larson, J. D. Joannopoulos, Phys. Rev. Lett, 68, 9 (1992)

[53] G. Binnig, H. Rohrer, Ch. Gerber, E. Weibel, Surf. Sci. 126 (1983) 236–244

[54] K. Takayanagi, Y. Tanishiro, M. Takahashi, and S. Takahashi, J. Vac. Sci. and Technol. A, 3, 3 (1985) 1502–1506

[55] R. Erlandsson, P. Apell, Current Science, 78, 12 (2000) 25

[56] A. N. Chaika, A. N. Myagkov, Journal of Physics: Conference Series 100 (2008) 012020

CHAPTER 4

EXPERIMENTAL FACILITIES AND PROCEDURES

4.1 Tip fabrication

All STM probe tips were homemade utilizing a W–Tek tip semiautomatic etching device from Omicron Nanotechnology, Germany [1]. A 0.3 mm in diameter tungsten wire placed at the centre of the cell (figure 4.1) was used as an anode, and a stainless steel wire loop was used as a counter electrode (cathode).

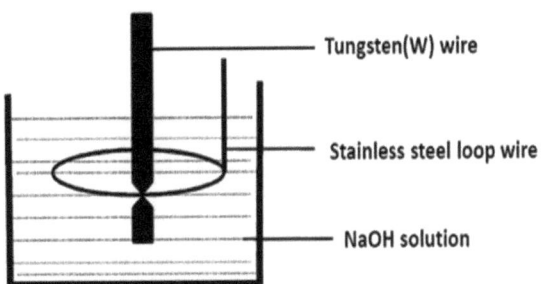

Figure 4.1. Schematic diagram of the electrochemical cell showing the W wire (anode) being etched in NaOH solution. The cathode consists of stainless steel wire loop which surrounds the anode (W).

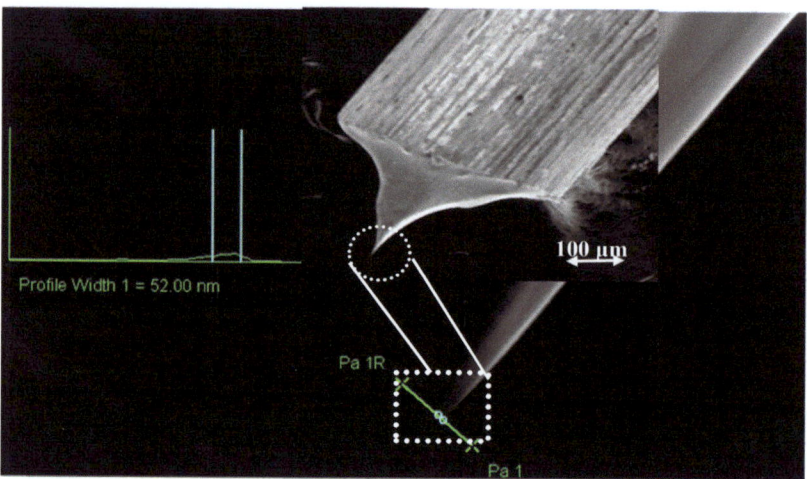

Figure 4.2. Scanning electron microscope image of atomically sharp STM probe tip after etching showing a tip diameter of ~52 nm.

The cathode was partially immersed into a 0.5 Molar sodium hydroxide (NaOH) solution using a micrometer drive to control the length of the wire under the surface of the solution. After the tungsten wire is aligned at the centre of the stainless steel wire loop, a DC current of a 3 mA is then applied through the tip to the solution to drive the etching reaction. The electrochemical reactions at the cathode and anode are as follows [2]:

$$\text{cathode: } 6H_2O + 6e^- \rightarrow 3H_{2(g)} + 6OH^-$$

$$\text{anode: } W_{(s)} + 8OH^- \rightarrow WO_4^{2-} + 4H_2O + 6e^-$$

(4.1)

At the anode, tungsten undergoes an oxidative dissolution of W to WO_4^2, which is soluble in water. At the cathode, water undergoes the reduction process to form bubbles of hydrogen gas and OH⁻ ions in equations 4.1. Because of these electrochemical reactions (4.1) the wire is etched at the contact areas with the solution thus forming a neck between the top part of the wire and the lower end. As etching progresses, the neck gets thinner and thinner and eventually breaks under the weight of the lower end of the wire leaving behind a sharp tip (figure 4.2). As soon as a sudden drop in current is detected by a feedback loop, an electronic circuit stops the process preventing over etching.

Alternatively one can use the AC self–terminating method, where the tip is inserted into the electrolyte and biased, generating a current that is constantly monitored by the feedback loop. In this method however, the wire material is completely etched away from the bottom end of the wire up to the wire electrolyte interface. As the wire cross–section is etched away, the current decreases until it reaches a pre–set value, at which time the bias is removed and etching presumably stops. It has been observed, however, that if the tip is left in contact with the electrolyte following etching, there is residual etching of the tip (and tip blunting), even without bias being applied and thus the AC method requires constant monitoring. Ultra sharp tips were obtained when using the previous mentioned method.

STM experiments are a time consuming process and generally it is good practice to assess the sharpness of the tips that are produced before performing experiments. The tip is removed from the solution, rinsed thoroughly with distilled water and dried with nitrogen, and then an optical microscope and scanning electron microscope is used to check if the tip is sharp enough for STM experiments. A tip considered suitable for STM experiments (figure 4.2) is transferred into the UHV system where it can be sharpened further by applying a strong electric field; this dislodges the surface atoms of the tip until only a few (ideally, one) remain. The SEM images show a limit of resolution of the instrument, however, in reality the diameter of the tip is much smaller than the one measures in SEM.

4.2 Characterization Techniques

4.2.1 The UHV variable temperature STM system

The Variable Temperature UHV Scanning Probe Microscopy system from Omicron, Germany, GmbH (figure 4.3) is composed of two parts: The main (experimental) chamber and fast entry chamber.

The experimental chamber is equipped with:

- Scanning Tunneling Microscope stage
- LEED–Auger electron spectrometer: operated by exploiting the same electron gun, grids and optics of a SPECTALEED tool.
- Sputter gun: sample sputtering with Ar+ ions operated with a ISE 10 sputter ion source.
- Heating/cooling sample manipulator. Molecular evaporator: organic molecules and materials with low sublimation point can be evaporated from a T–controlled Knudsen cell (up to a temperature of ~1200 K).
- Ion getter pump and titanium sublimation pump
- Cryostat for liquid helium bath

- Carousel with up to 6 sample position for storing sample plates and tip transfer plates.

The fast entry chamber includes:

- Magnetically activated sample transfer rod and is separated from the main chamber by means of a load–lock.
- Turbopump and roughing pump

Figure 4.3. The VT–UHV STM system from Omicron at the National Centre for NanoStructured Materials, CSIR

The UHV system can be cooled down to 25 K by filling a liquid Helium bath cryostat, which is then pumped by a diaphragm pump. Radiative heating can be achieved at the STM stage by using radiative heating sample plates which contains an embedded solid–state pyrolytic boron nitride (PBN) heating element that allows for temperatures up to 750 K. Direct heating with temperatures up to 1500 K is achieved via an electrical connection when the direct current heating sample plate is inserted into the sample acceptance stage.

Accurate temperature reading is not available for the commercial Omicron VT–STM, a major part of the development of the instrumentation is required. However to calibrate temperature measurements, the designers of the instrument measured the sample temperature when the thermocouple was attached directly on the sample plate and when the thermocouple was attached to its permanent position on the manipulator plate and the average of the measurements were recorded and gives a rough estimate to the sample surface temperature. One of the major issues regarding the VT STM is that it does not have facilities for accurate coverage measurements during growth such as a thickness monitor. The system relies on calibration curves which give a rough estimate of the thickness which has an error approximated to be 10%.

The sample is mounted in the STM sample stage with the surface pointing downwards and the tip is approached to the sample from the bottom up. The coarse approach, up to some tens of microns, is made using a CCD–camera. The fine approach is then performed automatically. Conversely to most STMs, in this instrument the sample is at ground potential and the bias is applied to the tip. Samples are introduced via the fast entry lock of the UHV system without breaking vacuum. The manipulator allow sample to be located at various positions inside the UHV chamber for sample preparations and analysis operations. The manipulator is equipped with sample heating facility (up to 1500 K) and a thermocouple for the sample temperature monitoring.

Substrate heating is achieved in two stages (direct or radiative heating). From the manipulator, the sample can be transferred to the STM stage by docking the sample plate on the sample acceptor stage in the UHV SPM, just above the protruding STM tip. Viewports on the bolt–on chamber allow optimal observation of the tip/sample coarse positioning using the external CCD camera.

4.2.2 Low–energy electron diffraction (LEED)

The surface of a crystalline solid in vacuum is generally defined as few (less than 10) outermost atomic layers of the solid that differ significantly from the bulk. It may be atomically clean or it may have foreign atoms deposited on it or incorporated on it. A complete characterization of a solid surface requires knowledge of not only what atoms are present but also where they are. Under suitable preparation conditions, the surface atoms of many materials are ordered and there are no restrictions to their movement to find their equilibrium positions. LEED is one of the most powerful techniques to determine the crystallographic quality of a surface, prepared either as a clean surface, or in connection with ordered adsorbate overlayers.

Analysis of LEED patterns and intensities provides the size and shape of the surface unit cell, the degree of order, and detailed atomic structure with a precision of the order of picometers. It may be used in one of two ways: qualitatively and quantitatively [3,5,8]. In qualitative analysis, the diffraction pattern is recorded and analysis of the spot positions yields information on the size, symmetry and rotational alignment of the adsorbate unit cell with respect to the substrate unit cell. In quantitative analysis, the intensities of the various diffracted beams are recorded as a function of the incident electron beam energy to generate so–called *I–V* (intensity vs. voltage) curves which, by comparison with theoretical curves, may provide accurate information on atomic positions. A typical experimental arrangement used in a LEED experiment is illustrated in figure 4.4. An electron gun is used to produce a well monochromatic electron beam [4].

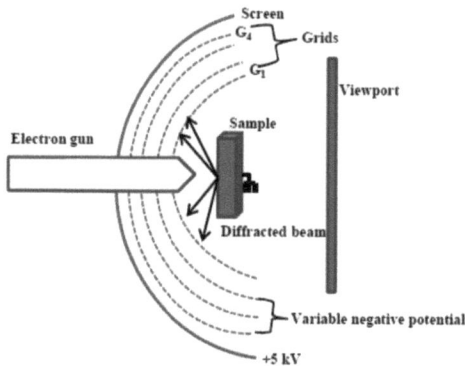

Figure 4.4. Schematic of the LEED optics.

The electron beam of energy 10 – 500 eV are incidents normally on the sample surface. These electrons are scattered mainly from the first few layers of the crystalline sample following the Bragg diffraction conditions. The backscattered electrons are of two types; elastically scattered electrons forming a set of diffracted beams which create the LEED pattern, and inelastically scattered electrons, which may make up 99% of the total flux, but they are not required. The de Broglie wave–lengths of these electrons are within 2.74 Å to 0.55 Å, which is optimal for crystallographic studies on the surfaces. Moreover for electrons within this energy range, the inelastic mean free path is 5 Å to 10 Å, which is equivalent to about 3–5 atomic layers [5]. Hence LEED is an excellent surface sensitive tool for crystallographic studies.

After reaching the first retarding grid G_1, which is earthed, the elastically scattered electrons are accelerated towards the fluorescent screen, which carries a high positive potential (of the order of 5 – 6 kV). This provides the electrons in the diffracted beams with enough energy to excite the fluorescence in the screen, so that a pattern of bright LEED spots corresponding to particular diffraction beams and actual images of the reciprocal lattice is seen. The next two grids (G_2 and G_3) are electrically coupled and are negatively biased at a potential slightly below the electron beam energy. These grids represent the filtering section where only the quasi–elastically scattered electrons are allowed to pass. The fourth grid (G_4) is also grounded to provide a field free region for post–diffraction acceleration. The potential on these grids is adjusted to minimize the diffuse background to the LEED pattern. The diffracted spots will move towards the centre of the image and higher order diffracted beams come into view as the primary energy of the incident electrons is increased. The interpretation of the changes in position and the intensity of the spots as a function of incident electron energy reveal much about surface reconstructions. For a more extensive review of diffraction techniques and LEED, the reader is directed to several texts [5–7].

4.2.3 Auger electron spectroscopy (AES)

The process of obtaining detailed information on the surface of a material is without doubt the most important and fundamental practice necessary for analysis in surface science.

Auger electron spectroscopy (AES) is used predominantly to check elemental composition of a freshly prepared surface with good spatial resolution under UHV conditions [3,9]. AES is surface sensitive due to the strong inelastic scattering which occurs at the energies of interest (~ 50–3K eV). Electrons with these energies originate near the surface or there is a high probability that the scattering processes will deplete them of the energy they require to escape into vacuum.

The process of creating an Auger electron starts when a primary electron with sufficient energy from an electron beam produces a hole in an inner shell of an atom in the sample (atomic ionization) such as the K shell (figure 4.5). When an electron from a higher shell (L_1) falls to fill the hole in the inner shell, energy is released. The energy may be enough to allow an electron in an outer shell (L_2) to leave the atom, and hereby become an Auger electron (Electron emission). This excitation process is denoted as a KL_1L_2 Auger transition and is shown in figure 4.5. The final stage of the spectroscopy entails the detection of Auger electrons with high sensitivity, and determining their kinetic energies utilizing a lock–in amplifier and energy analyzer respectively.

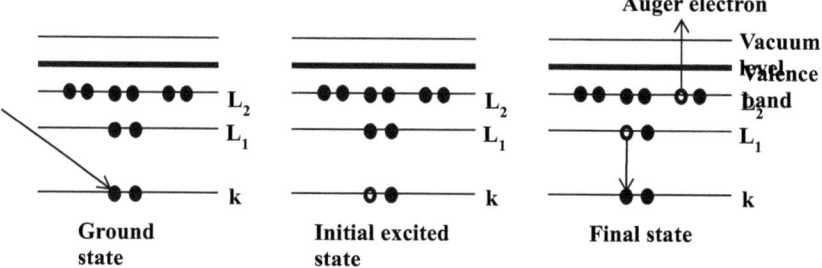

Figure 4.5. Energy level diagram of an Auger process. Electron from L1 drops into the K level with the emission of an L_2 electron.

The Auger electron will have energy given by

$$E = EK - EL_1 - EL_2. \qquad (4.2)$$

It is obvious that at least two energy states and three electrons must take part in an Auger process. This is why H and He atoms cannot give rise to Auger electrons. Several transitions (KL_1L_1, KL_1L_2, LM_1M_2, etc.) can occur with various transition probabilities. The energy of the Auger electron is characteristic of the parent atom and independent of the incident beam energy. Therefore, measurement of Auger electron energies constitutes a method of elemental identification. As a result of small Auger signals AES is usually carried out in the derivative mode to suppress the large background of the true secondary electrons [3,4]. The differentiation is performed by superimposing a small alternating voltage on the outer cylinder voltage and synchronously detecting the inphase signal from the electron multiplier with a lock–in amplifier. In this mode the detector current contains the first derivative dI/dV as the prefactor of the phase–sensitively detected AC signal with angular frequency. Auger line energies are usually given in reference works as the position of the minimum of the derivative spectrum dN/dE [3].

4.3 References

[1] Omicron NanoTechnology, Tip Etching Kit User Manual, Version 1.1 (1998)

[2] C. J. Chen, *Introduction to Scanning Tunneling Microscopy*. New York: Oxford University Press (1993)

[3] Y. W. Chung, *Practical guide to surface science and spectroscopy*, Academic Press, San Diego, CA (2001)

[4] D. P. Woodruff and T. A. Delchar, *Modern Techniques of Surface Science* (2nd ed.) Cambridge University Press (1994)

[5] R. A. Alberty and R. J. Silbey, *Physical Chemistry* (1st ed.) John Wiley & Sons, Inc., New York (1992)

[6] M. A. Van Hove, W. H. Weinberg and C. M. Chan, *Low Energy Electron Diffraction, Srpinger*, Berlin (1986)

[7] C. Kittel, *Introduction to Solid State Physics*, John Wiley & Sons, New York, NY (1996)

[8] J. B. Pendry, *Low–Energy Electron Diffraction,* Academic Press, London (1974)

[9] G. Ertl and J. Kuppers, *Low Energy Electrons and Surface Chemistry* (2nd ed.) VCH (1985)

CHAPTER 5

DISSOLUTION OF SB ON CU(111)

5.1 Introduction

For miscible metals, when one kind grows on the other's surface, the atoms near the interfaces will interdiffuse and form an alloy, which will affect the properties of the grown film. More interestingly, for materials with immiscibility less than 2 at % such as metal X (X= Bi, Pb, Sb) on Ag(111) or Cu(111), a kind of surface alloy formation is found under some conditions where every third substrate atom in the topmost layer is replaced by an alloy X atom resulting in a $(\sqrt{3} \times \sqrt{3})$ $R30°$–X structure [1–4]. As illustrated on the Cu–Sb binary phase diagram (figure 5), the miscibility of Sb in the bulk is ~ 1 at%. These systems of surface alloys between heavy metals and noble metals were recently identified by Ast et al. [5] as a new class of materials having long range order and exhibit large spin splitting of their surface states. The unusual embedded substitutional structure for an adsorbate–metal structure involving larger Z adatoms was already proposed years ago for Te (Z=52) on Cu(111) and Cu(100) [6]. While the existence of various surface phases has been widely recognized [7–14], the number of quantitative structure determinations is actually quite small.

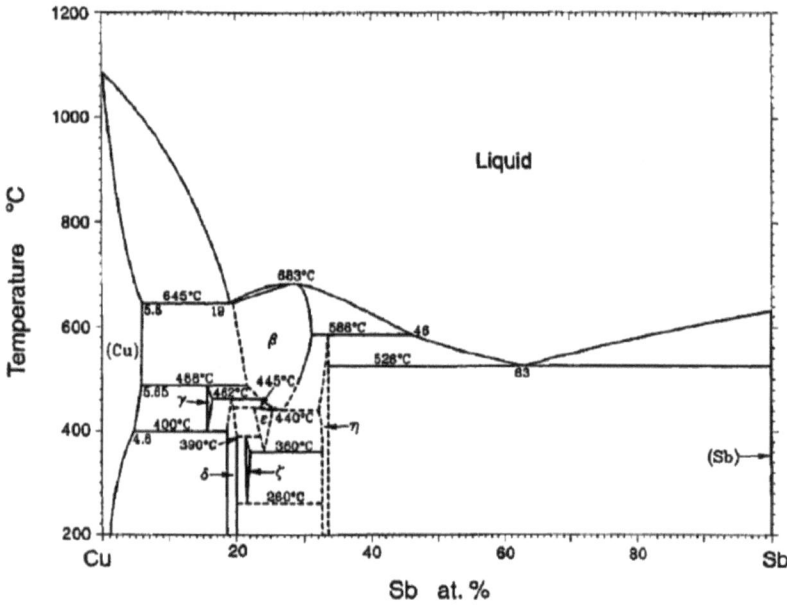

Figure 5. Binary alloy phase diagram of Cu and Sb in at%. Adapted from Massalski [15]

The system chosen for this study is antimony (Sb) on Cu(111), which is investigated in detail *in situ* using UHV variable temperature scanning tunneling microscopy (VT–STM) and its subsystems. Figure 5 depicts the binary phase diagram for copper and antimony. The Cu(111) – Sb system is quite analogous to the Ag(111) – Sb in a manner that both system present a tendency to phase separate, a strong surface segregation tendency of antimony (Sb) and lattice mismatch effect which causes rippling of the adsorbate atoms slightly above the substrate atoms [7–13]. The solubility of Sb in the bulk of copper (Cu) is limited by an intermetalic compound 3D–Cu_2Sb which is analogous for Sb/Ag(111) system [12,13]. Sb was chosen as the surfactant and employed at coverages of < 1 ML, which is the critical coverage for the Sb to induce a stacking fault in the Cu(111) surface [1–6]. The surface coverage is defined as the ratio between number of surface sites occupied by adsorbates and the total number of surface sites, and the corresponding unit is the monolayer (ML).

5.2 Experimental procedure

High purity (99.99%) copper single crystals with orientation accuracy of ~ 1° of the crystallographic (111) plane with 2 mm thickness from MaTeck, Germany, were used as substrates. The substrates were mounted on a Tantalum base plate with the polished side on top and introduced to the UHV system via a fast entry lock without breaking vacuum. To remove contaminants (sulphur, oxygen and carbon), the substrate was cleaned by cycles of sputtering and subsequent annealing in UHV as follows:

1. The Cu sample was sputtered with Ar^+ ions at 2 keV at room temperature for 5 minutes
2. The sample was annealed at 550°C and sputtered again for 5 minutes at 2 keV
3. The sample was further heated at 650°C for ~20 minutes without sputtering to level off sputter damage and concentration gradients.

Steps (2) and (3) were repeated four times until the impurity level was below the detection limit of AES and LEED indicated a sharp (1×1) pattern. Temperature was monitored using a chromel–alumel thermocouple modified such that the TC was in contact with the sample plate and very close to the crystal.

A Knudsen effusion cell was used for antimony (99.99% purity) growth at source temperature of ~455°C at a deposition rate of ~0.1ML/s. The chamber pressure was kept below 5×10^{-10} Torr during growth. The coverage of Sb was controlled by the evaporation time, measured deposition rates and crucible temperature charts supplied with the UHV system. The adsorbate coverage was calculated from acquired STM data after growth. The STM work was done in a UHV system (VT–STM, Omicron Nanotechnology, GmbH) equipped with sample characterization equipment such as LEED, AES optics, STM stage, effusion cell and sputter gun. LEED was utilized to study the sample surface before and after growth in reciprocal space. STM was used to determine ordered structural phases induced by Sb adsorption on the atomically clean Cu(111) surface. STM images were obtained with a low bias ranging from 0.3 to 0.001 volts in constant current mode in which the tip height is varied to keep the tunneling current constant while the tip is scanned laterally.

The STM data is displayed in a top–view with the intensity scale representing surface features, the darker levels corresponding to lower lying areas and brighter areas corresponding to higher lying areas. The effect of annealing the as–deposited surface was studied, and followed by subsequent annealing to progressively high temperatures followed by cooling to room temperature.

5.3 Results and Discussion

5.3.1 Atomically clean Cu(111)

Careful characterization of the clean copper (111) surface before deposition of Sb is a prerequisite in understanding the growth mechanism and behavior of the Cu(111) – Sb system. The STM data for all recorded images is taken at different scanning directions to exclude any tip–induced artifacts. In most STM images of metallic surfaces, the observed maxima are attributed to the effective atomic positions [14,15].

An STM image of atomically resolved hexagonal structure of Cu (111) surface is depicted in figure 5.1 (a) with nearest neighbor distance of 0.257 ± 0.01 nm (indicated by the line C in the topography image). This value compares well with tabulated values of 2.56 nm [16] for the atomically clean Cu(111). The variations are attributed to surface layer vibrations. The "intensity" scale next to the image illustrates the height of the features on the image ranging from 0 – 39.6 pm. The corresponding line profile (figure 5.1 (b)) taken along the line AB in figure 5.1 (a) depicts the corrugation of the Cu atoms to be ~ 0.024 nm on average.

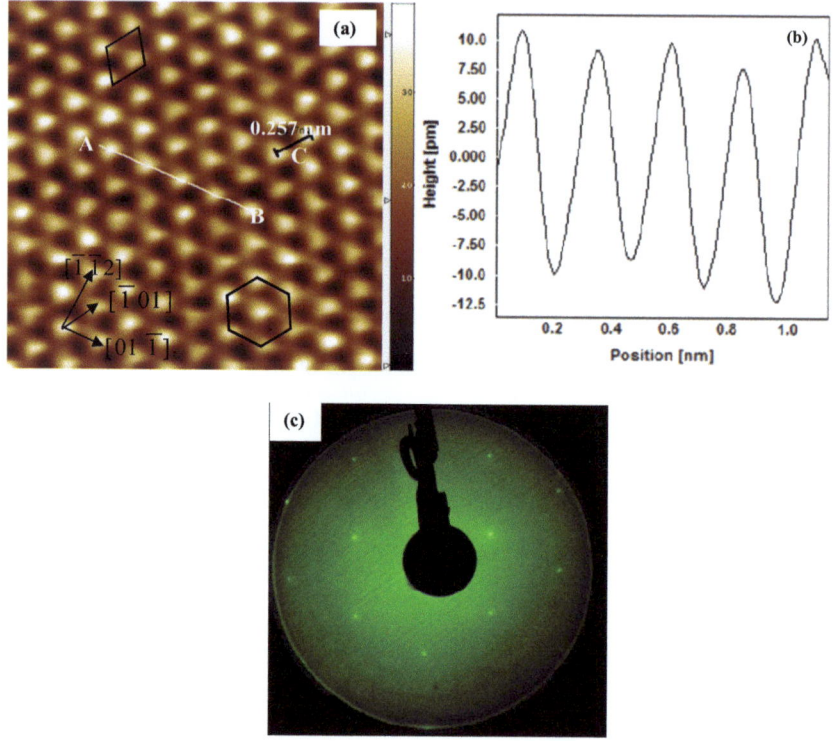

Figure 5.1. (a) High resolution STM image of atomically clean Cu(111) surface (2.50 nm × 2.50 nm) $V_{bias} = -1$ mV, $I_{tun} = 2$ nA. The black rhombus indicates the (1×1) unit cell and hexagon is superimposed to illustrate the arrangement of Cu atoms at the surface. The superimposed unit vectors on the topography image illustrate the directions on the FCC Cu surface. (b) Corresponding line scan along the line AB in (a). (c). LEED pattern of the bare Cu(111) surface recorded at 114 eV showing the (1×1) main spots.

The atomic spacing is measured from the top of the crest of a protrusion to nearest protrusion and is alternatively confirmed by a unit cell detection method where unit cells are superimposed on the entire STM image and atomic spacing are extrapolated from the unit vectors. This unit cell detection feature is provided by the commercially available scanning probe image processor (SPIP) software package. As expected, the LEED pattern (figure 5.1

(c)) of the bare copper surface depicts a typical (1×1) pattern characteristic of a well ordered (111) surface at a beam energy of 114 eV. The pattern illustrates bright spots with low background intensity. The hexagonal arrangement of Cu atoms in reciprocal space is readily seen from the pattern.

5.4 Growth of Sb on clean Cu(111) surfaces

The first step of the growth process is chemisorption of the atomic species on the surface of the substrate. Chemisorption is generally a simple process for atomic species. For molecular species, however, chemisorption is often far more complex, involving bond breaking and surface interactions between multiple molecules. When atoms from the gas phase come into contact with a metal surface the adsorbates experiences the periodic substrate atomic lattice mediated by both the surface energy of the substrate and the surfactant nature of the Sb adsorbates (change either the surface energy of the growing layer or growth kinetics). The binding energy of the adsorbates is subject to lateral variation with local minima corresponding to energetically favorable positions called the adsorption sites. As presented in **chapter 2**, real surfaces very often consist of mixtures of flat terraces presenting steps, kinks, and point defects. These adsorption sites are separated by energy barriers being significantly smaller than the energy barrier for desorption. The minimum energy difference between adjacent sites is the migration energy barrier E_m. The reported surface energy for low index copper orientations are: 0.707 eV per atom for (111), followed by 0.906 eV per atom for (100) and 1.23 eV per atom for (110) [23].

After growth, a thermodynamic interaction and redistribution of the near surface crystal layers establishes the final configuration. The high resolution STM image of figure 5.2 (a) was acquired after deposition of ~ 0.3 ML Sb on the atomically clean Cu(111) surface, at room temperature.

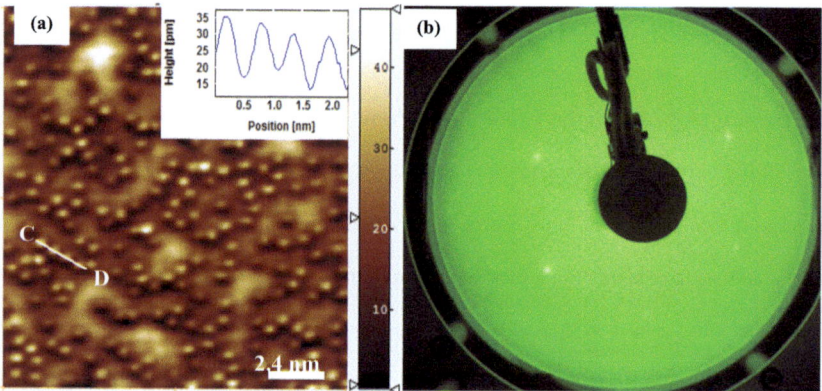

Figure 5.2. (a) Atomic resolution STM image upon deposition of ~0.43 ML of Sb (V_{bias} = 5 mV, I_{tun} = 4 nA). Adsorbate atoms substitute into top layer Cu lattice sites and are imaged as protrusions. Inset in (a) shows corresponding line scan taken along the line in (a) showing the corrugation of Sb adatoms on copper. (b) LEED pattern captured at beam energy of 116 eV.

Figure 5.2 (a) show bright spots (protrusions) which can be associated with Sb atoms on the Cu surface. The Sb atoms are more or less randomly distributed and embedded on the Cu surface. The inset on figure 5.2 (a) illustrates the corrugations of the Sb atoms on the Cu surface along the line C to D. As the Sb atoms leave the source they are at high temperature ~ 450°C and are incident on a substrate kept at room temperature. Upon landing on the Cu surface, the Sb adsorbate atoms interact strongly with the Cu surface, showing a strong tendency to wet the surface due to the surfactant nature of Sb (figure 5.2 (a)). This type of growth mechanism (Frank van der Merwe) can be further understood by taking into account Young's equation defined as [24]

$$\gamma_s = \gamma_{as} + \gamma_a \cos\theta \qquad (6.1)$$

where γ_s, γ_{as} and γ_a are the surface free energy, interface free energy, film free energy respectively and θ is the contact angle between adsorbate and substrate. Following equation (6.1), it is possible to determine beforehand whether metal atoms or molecules are going to wet the surface or not. For a contact angle of 0 degrees, the metal atoms wet the surface and equation (6.1) becomes $\gamma_s = \gamma_{as} + \gamma_a$.

After growth $k_BT \ll E_m$ which means the Sb adsorbates are confined to the adsorption sites, i.e. the effect of the lateral Cu surface corrugation on the Sb adsorbate motion becomes large. However, before this equilibrium state, the Sb atoms must be sufficiently mobile to reach adsorption sites of minimum free energy through a diffusion process which follows Arrhenius behavior given by

$$D = D_0 \exp\left(\frac{-E_{act}}{RT}\right) \qquad (6.2)$$

where D is the diffusion coefficient, E_{act} is the activation energy barrier and D_0 is the pre-exponential factor called diffusivity. The site of Sb atoms remain stable for long period of time (minutes) during STM scans in contrast to the anticipated high diffusivity of single Sb adsorbate atoms on FCC surfaces. This might be due to the Sb atoms being embedded in the first Cu layer, forming a surface alloy with Cu(111) surface or alternatively, it could be due to the surfactant nature of Sb which means the surface reach thermodynamically equilibrium and lowest energy configuration in this manner.

A number of STM images similar to that of figure 5.2 (a) were used to calculate the surface Sb coverage after growth. This was done by selecting an area showing Sb atoms in atomic resolution from the STM images. From known atomic radius of copper atoms (and the atomic resolution data of bare Cu(111) in figure 5.1 (a)), the number of Cu atoms which can fully populate the selected area was determined. Using the scanning probe image processor (SPIP), the number of Sb atoms at the select area was acquired. Even though the Sb atoms are randomly distributed after growth, the Cu(111) surface retain its structure after as seen from the LEED patterns (figure 5.1 (c)), thus the ration of the Sb atoms with respect to the Cu substrate atoms was computed gave a value of 0.43 ± 0.02 ML. This value is a true representation of the surface coverage after growth and is within acceptable magnitude e when compared to the estimated 0.3 ML as measured from calibration tables.

The corresponding LEED studies gave a diffuse pattern (figure 5.2 (b)) which displayed no additional features to the (1×1) clean surface. LEED is an averaging technique and since Sb atoms are not arranged in a regular pattern within the Cu(111) surface, Sb atoms makes no contribute to the electron diffraction spots which is why only the Cu(111) surface atomic arrangement is captured by LEED. The presence of defects, in this case Sb atoms on the atomically clean Cu(111) surface results in broadening and weakening of the spots and an increase on the background intensity.

5.5 Stages of dissolution

5.5.1 Annealing at 360°C

The simplest reason for a contrast between different atom species at the surface is a true topographic effect, i.e., the difference in atom size and/or difference in atomic position (height). Discrimination between different metal atoms at many different alloy surfaces has been observed in constant current mode, a phenomenon known as chemical contrast [27]. To initiate the dissolution process, the Cu(111) – Sb surface was annealed at 360 ± 10°C for 12h as captured by STM in figure 5.3 (a). The array of bright dots in the image (marked B) are identified with the Sb atoms while the dull or less bright (marked D) and smaller dots correspond to isolated Cu atoms. At this annealing temperature, $k_BT \sim E_m$ to $k_BT \gg E_m$ which means the thermal energies are close to or exceeding the migration energy barrier and thus the effect of the lateral Cu surface corrugation on the Sb adsorbate motion becomes smaller or even negligible. Surface migration is thus less restricted and the adatoms transport or diffuse freely on the surface without confinement to specific sites.

In order to minimize the surface energy of the Cu(111), the Sb atoms re–arrange themselves in an almost regular fashion which exhibits long range order. This effect is demonstrated by the rearrangement of Sb atoms from the previously random distribution after growth (figure 5.2 (a)) to an arrangement which is almost hexagonal on the Cu(111) surface upon annealing (figure 5.3 (a)). The line profile (figure 5.3 (b)) taken along the line A to B in figure 5.3 (a) depicts the corrugation of both the Cu (green and turquoise markers) and Sb (blue and red markers) atoms.

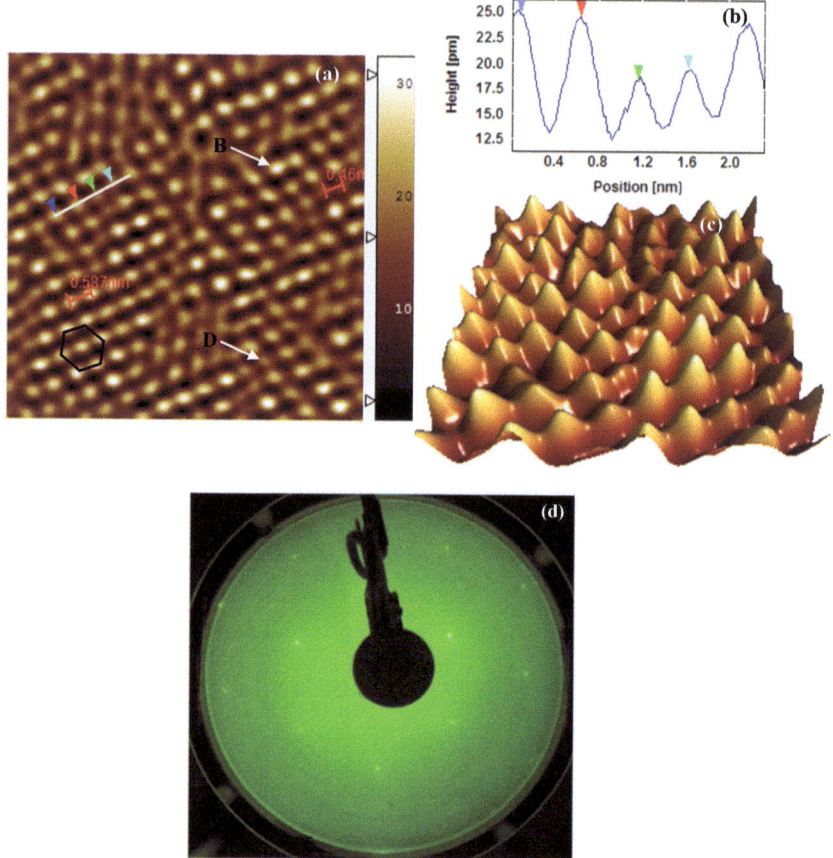

Figure 5.3. (a) Atomically resolved STM image of Sb atoms incorporated in a Cu(111) after annealing at 360°C (8.74 nm × 8.30 nm). The image was taken at V_{bias} = 1 mV, I_{tun} = 2 nA tunneling conditions. Side bar show height in pm. (b) Line scan taken along the line in (a). (c) 3D view of the image in (a). showing the corrugations of both Sb and Cu substrate atoms. (d) LEED show Cu(111) underlying structure since the Cu is still clean and has a regular pattern than Sb so we see Cu pattern.

The spacing between antimony atoms was measured to be ~0.586 nm on average and Cu nearest neighbour was 0.461 nm as seen on figure 5.3 (a–b) which is almost double the spacing on clean Cu(111). This increase in bond length of Cu atoms illustrates that the entire surface (both Cu and Sb atoms) has undergone surface reconstruction during the annealing process. Figure 5.3 (c) depicts a 3–dimensional view of the far left portion of figure 5.3 (a) where Cu atoms appear smaller and lower than the Sb atoms. The variation in corrugation and size of the Sb and Cu atoms is self–evident from the line profile in figure 5.3 (b). The degree of rippling at the surface can be expressed in terms of the effective radii of the Sb atomic species accommodated into the surface of the Cu substrate. The STM observations suggest that surface alloy formation is favourable even at 360°C. Although the surface has not completely reconstructed to a complete recognizable structural phase, LEED exhibits an average of the contributions of all domains existing on a surface in this case a sharp and well contrasted pattern with no extra spots of a hexagonal closed packed surface (figure 5.3 (d)).

5.5.2 Annealing at 400°C

Upon increasing the sample temperature, the sample surface may start to melt at temperatures well below the bulk melting point of Cu ~ 1084°C. It is known that for Cu, surface adatoms begins to appear on the surface above 530°C which eventual results to the onset of surface premelting (the localized loss of crystalline order at surfaces and defects at temperatures below the bulk melting transition) with planar disorder at and above 930°C [29–32]. In this case the annealing temperature is way below the roughening or the premelting temperature, thus the influence of pre–melting is excluded from experimental observations. A careful analysis of the CuSb sample after annealing at 400°C for 12 hours indicate the dissolution of Sb and showed a $(\sqrt{3} \times \sqrt{3})\,R30°$–Sb superstructure (figure 5.4 (a–b)).

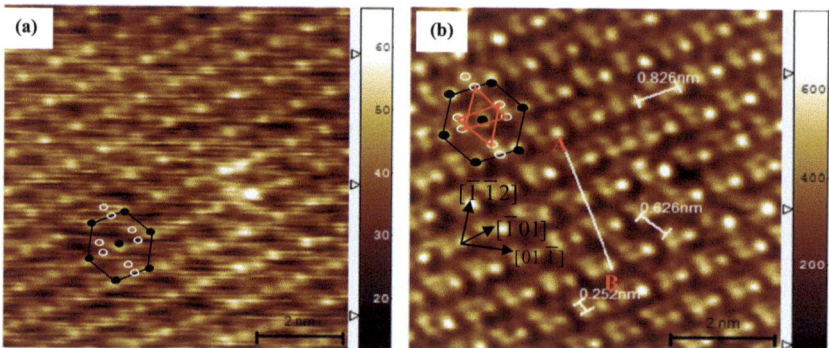

Figure 5.4. (a–b) STM images clearly showing the six Cu atoms that surrounds each Sb atom in a Cu(111) –$(\sqrt{3} \times \sqrt{3})\,R30°$–Sb reconstruction ($V_{bias}$ = 1 mV, I_{tun} = 2 nA). The intensity scale in (a) represent tunnelling current variations in pA while (b) show variations of height in pm. The superimposed hexagon is with respect to the Sb atoms (black dots) and the open white circles represent Cu atoms. The superimposed triangles are associated to two in–equivalent Cu atoms and a Sb atom at the centre resulting in surface asymmetry in (b).

At V_{bias} = 1 mV, I_{tun} = 2 nA tunneling conditions, both Sb and Cu atoms are simultaneously imaged by STM. The vertical (z) corrugations of figure 5.4 (a–b) represent height variations in pm and tunneling current in pA respectively. A closer look to figure 5.4 (b) it is possible to discern that Sb atoms occupy substitutional sites surrounded by six Cu atoms, three of them being displaced from the lattice site positions of the (111) plane. This indicates that the substitutional character of surface segregation is a general result for metallic systems, independent of the atomic radii mismatch between Sb and Cu. The Sb coverage was determined to be more that the 0.3 ML which is the critical coverage to induce the conventional ($\sqrt{3} \times \sqrt{3}$) $R30°$–Sb superstructure, thus the annealing temperature of 400°C is sufficient to drive off any excess Sb to the subsurface and produce an ordered surface phase.

The Sb atoms adopt the symmetry of the substrate Cu (FCC) atoms. The spacings between the Sb atoms show an increase from 0.587 nm as seen from the initial stages of dissolution (figure 5.3 (a)) to ~ 0.826 nm (figure 5.4 (b)). The increase in temperature results in the increase in the diffusion of the atomic species on the sample surface which later reach thermal equilibrium but not after the bonds holding the atoms together are broken and the surface rearrange to a stable configuration with the lowest energy possible. The process of surface reconstruction is thus directly responsible for increase in bond length. The two sets of triangles in figure 5.4 (b) of Cu atoms along the <112>–type direction illustrates the symmetry breaking of the (111) surface. This observations rules out the proposed model (see **section 2.5**) of Sb atom sitting as adatoms on the Cu surface due to the large difference in atomic radius between Cu and Sb.

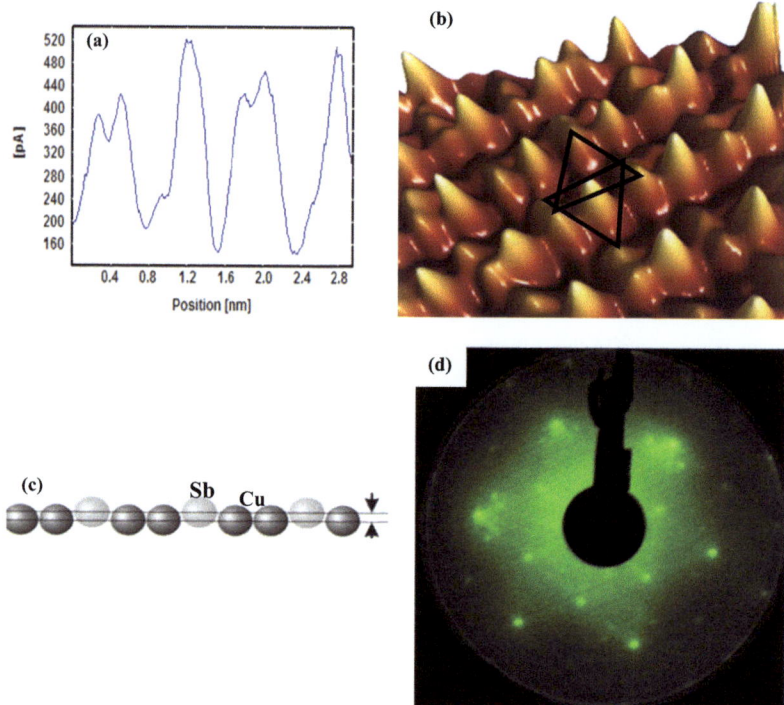

Figure 5.5. (a) Line trace a long A to B in figure 5.4 (a) showing the amplitude of the corrugation in pico–amps vs distance in nm. (b) Three–dimensional (3D) view of the high magnification STM image of the image in figure 5.4 (b). The superimposed star is associated to two in–equivalent Cu atoms and a Sb atom at the centre resulting in surface asymmetry. (c) Schematic representation of the CuSb atomic arrangement at the surface. (d) LEED pattern of the sample surface at 114 eV illustrating the $(\sqrt{3} \times \sqrt{3})R30°$–Sb superstructure.

It can be noted that the rearrangement of Sb atoms to occupy substitutional sites on the surface alloy is unusual, but not unprecedented [7–9]. A similar surface alloy formation was observed in previous studies of Sb/Ag(111) [2,9] (lattice mismatch ~1 %) indicating that the main driving force for the two dimensional (2D) surface alloy formation is due to the tendency of the materials to chemically order which is almost independent of the lattice mismatch between Cu substrate and Sb deposit. It is known that perfectly symmetrical structures can be distorted towards less symmetric, but more stable ones in order to minimize their energy by enhancing the weight of the bonding states in the LDOS [10,19]. It is thus appealing to interpret the contrast observed on the acquired STM images as being due to the reconstruction of the surface layer in which three Cu atoms forming a triangle "go up" whereas the other three "go down" i.e., the Cu atoms are shifted from their equilibrium positions. The Cu(111) – Sb system have undergone a similar process which results in the observations of figure 5.4 (a–d). From the STM images (figure 5.4 (a–b)) the Sb atoms remain centered in the <112> –type directions and followed by two lower Cu atoms evident in the line scan of figure 5.5 (a). The out of plane displacement (rippling) of Sb incorporated into the copper surface is consistent with the picture of the larger antimony atoms substituting for copper as illustrated by the 3D view in figure 5.5 (b) and the schematic in figure 5.5 (c).

The corresponding corrugation shown in figure 5.5 (a) is strong but it should be kept in mind that what is measured here is the corrugation of the surface charge density which can be enhanced with respect to the atomic one. The inclusion of the Sb in the Cu matrix induces strain in the Cu layer which leads to a large distortion within the Cu top layer chains. The strain is reduced by the lateral shifting of the Cu atoms which is evident from the outward relaxation of the surface atoms (figure 5.5 (b)). Previous studies using density functional theory (DFT) method by Woodruff and Robinson [9] found that for $(\sqrt{3} \times \sqrt{3})$ *R30°*–Sb periodicity, the stacking–faulted configuration is indeed more stable than an unfaulted configuration. Their obtained structural parameters were in excellent agreements with the experimental data obtained by Bailey et al. [8] and Umezawa et al. [17]. In a similar study de Vries et al. [2] reported that at a coverage of 0.33 ML Sb, the top layer atoms reside at hexagonal closed packed hollow sites relative to the underlying Cu(111) surface, while for coverages below 1/3 ML, the Sb atoms are embedded randomly at FCC positions in the top surface layer. From the acquired STM data in this study, Sb atoms buckle outward by ~0.23 nm (i.e. the extent to which the constituent atoms of the alloy layer are not strictly coplanar) compared to the previous reported theoretical and calculated values of 0.25 nm [18]. The outward relaxation of the alloy atoms in these surface alloys using quantitative low-energy electron diffraction (LEED) was determined previously by Gierz et al. [1]. The hexagonal surface symmetry is then broken by this displacement as illustrated in the three–dimensional picture in figure 5.5 (b).

The $(\sqrt{3} \times \sqrt{3})R30°$–Sb structure was investigated by LEED and the corresponding LEED pattern (figure 5.5 (d)) presents the $(\sqrt{3} \times \sqrt{3})R30°$–Sb reconstruction corresponding to 1/3 ML Sb when taken at 114 eV. Extra spots (superspots) are visible on the diffraction pattern compared to the clean Cu surface of figure 5.1 (c) showing only the integer–order or main spots. This arrangement of surface atoms can be understood by considering that, in vacuum, the valence electron density of metal surfaces is reduced by spilling–over and results in tensile surface stress which is relived by allowing large substitutional Sb atoms to be accommodated on the Cu substrate surface. It is rather intricate to be more quantitative from the analysis of the STM images, thus, a detailed structural analysis using the spectroscopic part of the STM is desired.

5.5.3 Annealing at 600°C

The sample was further annealed at 600°C for 12 hours. STM scans were taken at a sample temperature of ~250°C and captured somewhat more complex super structure (figure 5.6 (a)) which can be explained by both experimental and theoretical models. The STM data provides some insight into the ordering behavior of the Sb atoms in the surface after annealing. The intense spots define the overall periodicity, as indicated by the unit cell drawn as a black rhombus in figure 5.6 (a). The degree of ordering is strongly dependent on both the composition of adsorbate atoms and annealing temperature.

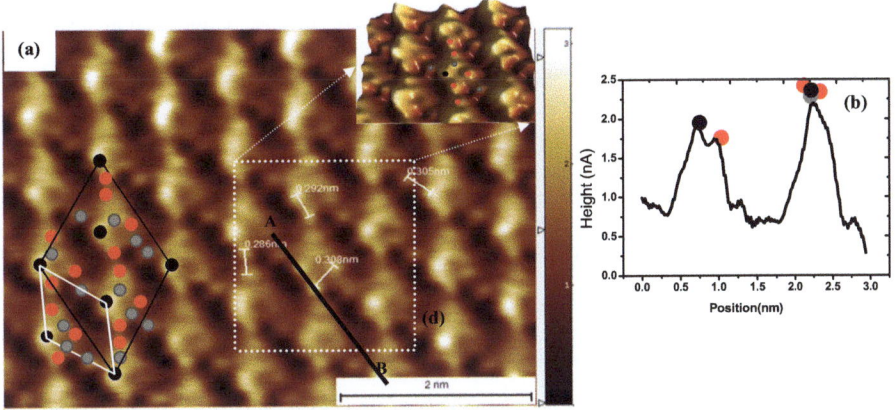

Figure 5.6. (a) Fourier filtered STM image showing the $(2\sqrt{3} \times 2\sqrt{3})R30°$ superstructure. The $(2\sqrt{3} \times 2\sqrt{3})$ unit cell is depicted by the black rhombus. The superimposed black dots represent positions of the overlayer Sb atoms, while the intensity dots represent the substitutional Sb atoms and the red dots are associated with copper atoms. Tunneling conditions: $V_{bias}= 1$ mV, $I_{run} = 3$ nA. Both the Sb and Cu atoms are resolved and the black rhombus outlines the $(2\sqrt{3} \times 2\sqrt{3})R30°$ unit cell while the white rhombus depicts the (2×2) sub–cell. The inset show 3D view of the image. (b) Line scan taken along the black line AB in (a) showing the corrugations of both Sb and Cu atoms.

In this study, it is proposed that, the atomically resolved STM images (figure 5.6 (a)) reveals two types of antimony atoms at the Cu(111) sample surface forming a ($2\sqrt{3} \times 2\sqrt{3}$) *R30°*–Sb superstructure. The variation in LDOS can be used to differentiate between different atoms at the surface. LDOS contrast includes cases where one type of atom is significantly larger and/or geometrically higher than the other and, hence, imaged higher by STM. The difference in size (metallic radii) of Sb and Cu is large ~ 15 pm in radius. With such a large size difference, it has to be expected that the Sb atoms relax outwards from the Cu surface layer. The overlayer Sb atoms which sits on the HCP sites of the ($\sqrt{3} \times \sqrt{3}$) *R30°*–Sb surface alloy appear with brighter contrast on the STM image as observed after annealin. The other set of Sb atoms are the substitutional Sb atoms on the Cu(111) plane which are represented by grey circles in figure 5.6 (a).

Furthermore, depicted in figure 5.6 (b) is the corrugation of both Sb and Cu atoms taken along the line A to B in figure 5.6 (a). The corrugation of the overlayer Sb atom with the substitutional Sb and copper atoms are imaged as individual entities by the STM. The effect of imaging the molecule as a single protrusion is seen through the variation in height of the real overlayer atom and the molecule. Close inspection of the image in figure 5.6 (a) reveals that the evidently large measured distances (~0.83 nm) between the bright spots as well as the different symmetry of the local areas near each bright spot prove that the corrugation maxima in the STM images cannot be assigned to single antimony atoms in a lattice plane of Cu. However a closer look after filtering of the image show clearly the atomic positions of different atoms (inset on figure 5.5 (6)). From this observation it is evident that the large lattice mismatch between Cu and Sb does not necessarily hinder the formation of a surface alloy even at high temperatures. Each overlayer Sb atom in a lattice plane has about six next neighbours at a various distances as illustrated by the measured distances of figure 5.6 (a). Starting from the measured distances between the corrugation maxima in the local, well–ordered areas of the over layer Sb atoms (0.832 nm), the ordered antimony structures can hence be explained, if each protrusion measured with STM, (i.e., maxima of the local electronic density of states), is assigned to one undissociated Sb atoms sitting on top of two Cu atoms and a substitutional Sb atom thus forming a tetrahedron molecule. Sb is known to form tetrahedrons at temperatures above 555°C [25,26]. The tetrahedron molecule is well resolved in the 3D image which is an inset on figure 5.6 (a) taken on the marked white rectangle. Taking into consideration the convolution with the STM tip, this arrangement of atoms represents an upper limit for the lateral size of Sb and thus agrees well with the observed contrast on the STM images.

Figure 5.7 (a) illustrates the proposed schematic representation of the real–space structural model of the surface alloy. The model outlines an overlayer of Sb atoms (black circles in figure 5.7 (a)) superimposed on an Sb–substituted $(\sqrt{3} \times \sqrt{3})$ $R30°$–Sb surface with a periodicity corresponding to a p(2×2) phase relative to the Cu bulk substrate. The preference for the formation of a Sb p(2×2) structure is highlighted by the proposed structural model.

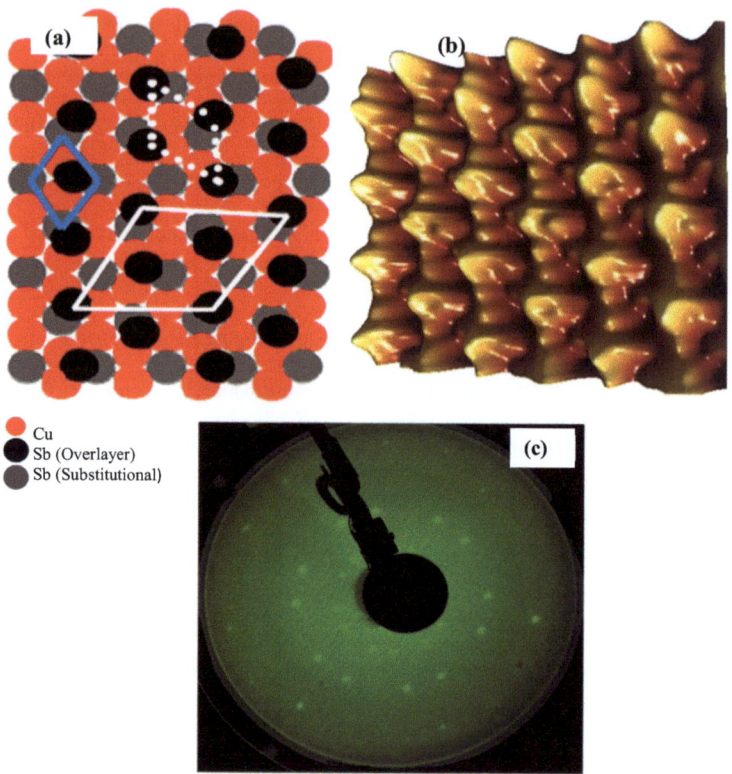

Figure 5.7. (a) Atomic model for the $(2\sqrt{3} \times 2\sqrt{3})\,R30°$–Sb superstructure outlining Sb atoms adsorbed in FCC (or HCP) hollow sites above a substitutional $(\sqrt{3} \times \sqrt{3})\,R30°$–Sb structure. The $(\sqrt{3} \times \sqrt{3})\,R30°$, (2×2) and $(2\sqrt{3} \times 2\sqrt{3})\,R30°$ unit cells are outlined. (b) Fourier filtered 3Dview of STM image in (a) showing more details of the atomic positions. (c) Corresponding LEED patter taken at 200°C at beam energy of 114 eV.

From a theoretical standpoint, the traditional approach for the determination of an alloy structure implies, in principle, a search through any possible configuration until the most energetically favourable is found. From the proposed structural model (figure 5.7 (a)), the most likely coverage of Sb atoms is three to five additional Sb atoms per $2\sqrt{3}$ unit mesh relative to the $\sqrt{3}$ starting structure. Thus the proposed Sb coverage of the surface as acquired by STM is ~0.43 ML. The model show overlayer Sb atoms to occupy identical sites, bonding to two Cu atoms and one substitutional Sb atom in a set of hollow sites in the substrate. From both the experimental and the proposed model, it can be seen that the new structure which give rise to the p(2×2) reconstruction penetrates deeper into the copper substrate. It can be seen from the model that the p(2×2) unit cell is consistent with Cu_3Sb stoichiometry for this alloy phase.

The symmetry of the surface and its structures is particularly well resolved as represented by the 3D view images of figure 5.7 (b). The surface reconstruction of the Cu(111) – Sb system depicts $(2\sqrt{3} \times 2\sqrt{3})R30°$ –Sb structure showing overlayer Sb atoms surrounded by Cu atoms together with substitutional Sb atoms. A similar $(2\sqrt{3} \times 2\sqrt{3})R30°$ structure was obtained for Sb/Ag(111) and In/Cu(111) systems [3,20,28] at ~ 0.6 ML coverage, but has not been reported for Cu(111) – Sb system.

The arguments given previously in explaining the $(\sqrt{3} \times \sqrt{3})R30°$ –Sb structure that Sb enters the Cu surface substitutional is still applicable on the observed $(2\sqrt{3} \times 2\sqrt{3})R30°$ – Sb structure. The sample surface shows a honeycomb structure and still retains the $(\sqrt{3} \times \sqrt{3})R30°$ periodicity with a substantial displacement of the subsurface atoms. To differentiate between different atoms from the acquired STM images, the proposed model of figure 5.7 (a) is taken into consideration and serves as a guide for atomic position at the sample surface. It is also interesting to note that, in spite of the higher strain induced by larger Sb atoms inside the Cu substrate, the alloying effect of the strong Sb–Cu bonds results in the net effect of Sb interdiffusion leading to the formation of a Cu_3Sb phase in the near-surface layers or alternatively Cu atoms diffuses upwards to form Cu_3Sb as in the case of In/Cu(111) system [28].

The Cu–Cu distance is measured to be ~ 0.287 nm on average with rippling value of ~ 0.02 nm with respect to substitutional Sb atoms. The mean distance of the Sb overlayer protrusions (intense spots) as measured from the STM images is ~0.831 nm on average. This Sb–Sb atom distance is comparable to the Sb–Sb distance measured on the $(\sqrt{3} \times \sqrt{3})$ $R30°$–Sb structure of the surface alloy at 400°C (figure 5.4 (b)). The lack of alternate ordering patterns that compete energetically with the one observed in the STM data above, leads to the conclusion that for at higher temperatures, the system reconfigure its self to an energetically metastable configuration which amounts to a suppressed number of Cu atoms at the surface as observed experimentally. The STM data show both the Sb and Cu atoms at the surface however the observed surface structure is incommensurate with the substrate lattice which makes linking the experimental data to the model intricate.

To further investigate the metastable Cu(111) $(2\sqrt{3} \times 2\sqrt{3})$ $R30°$–Sb phase, LEED patterns were taken and showed a $(2\sqrt{3} \times 2\sqrt{3})$ $R30°$–Sb structure (figure 5.7 (c)). The LEED pattern of the surface taken at 114 eV showed a ring around the (0.0) beam spot characteristic of an ordered $(2\sqrt{3} \times 2\sqrt{3})$ $R30°$–Sb structure. The ring–shaped diffraction pattern indicates a uniform Sb adatom distribution on the Cu surface [21]. The two rings composed by twelve equidistant spots are consistent with the growth of a hexagonal film forming three equivalent rotational domains [22]. Increasing the LEED electron energy up to 360 eV showed no influence to the (2×2) pattern, suggesting that the reconstruction persist into deeper layers below the surface implying interdiffusion. Once the p (2×2) pattern has appeared, no further changes on the diffraction pattern were observed when the sample temperature was increased. This indicates that a Cu(111)p(2×2)Sb structure is formed and that no more than the proposed number of Sb atoms stays on the substrate even at higher temperatures. Upon cooling to room temperature the LEED pattern reverted to a $(\sqrt{3} \times \sqrt{3})R30°$ pattern which implies that the Sb concentration dropped to that of the $(\sqrt{3} \times \sqrt{3})R30°$ structure and the adsorbed atoms change sites during this phase transition.

5.5.4 Annealing at 700°C

Figure 5.8 (a) displays an STM image of sample surface after annealing at 700°C. After a careful examination of the STM image and the corresponding line profile (figure 5.8 (b)), it is possible to distinguish between the atomic positions of the Sb and Cu atoms. The interatomic spacing between the Sb atoms is 0.810±0.002 nm which is comparable to the measured spacings between Sb of the previous reported superstructures at 400 and 600°C.

The STM data reveals only four Cu atoms instead of the six surrounding each Sb atom (figure 5.8 (a)). This implies that each hexagonal arrangement of Sb atoms has two atoms less than in $(\sqrt{3} \times \sqrt{3})R30°$ and thus the Sb surface concentration at this elevated temperature has reduced from 0.3 ML. The Cu atoms are centered in the <110> – type directions.

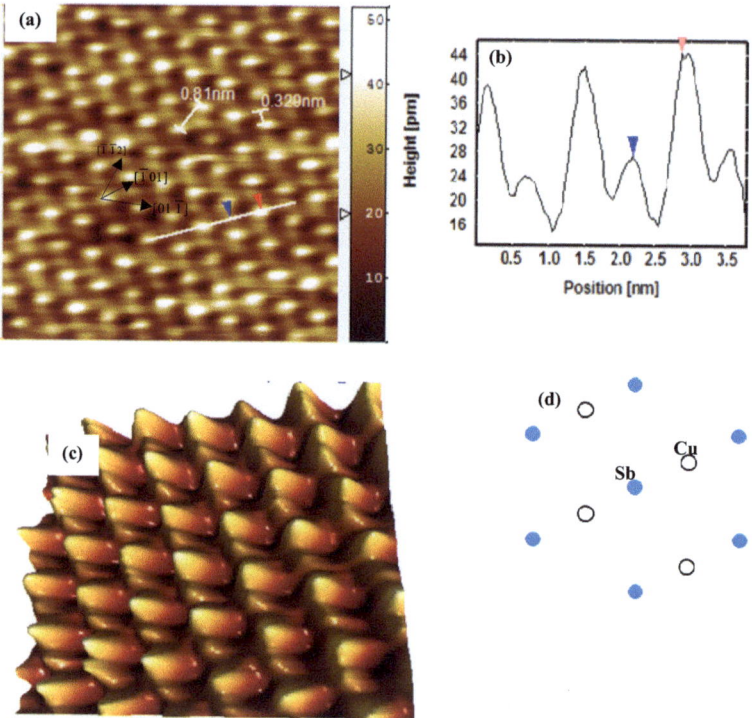

Figure 5.8 (a) STM image (Vbias = 0.7mV, I = 4 nA) of Sb dissolved at the Cu (111) surface after annealing at ~700°C forming hexagonal arrangement of a ($\sqrt{3} \times \sqrt{3}$) $R30°$ + (2×1), (b) line profile taken along the line in (a), (c) three–dimensional view of the STM image in (a) and (d) Schematic illustration of the positions of both the Sb and Cu atoms.

The line scan in figure 5.8 (b) taken along the line in figure 5.8 (a) depicts the relative corrugations between the Sb and Cu atoms as represented by the colour coding of arrows on both the image and the line scan. The average distance between Cu–Cu atoms is 0.757 nm. The STM data still maintain the rippling effect (0.02 nm) observed on the previous ($\sqrt{3} \times \sqrt{3}$) $R30°$ reconstruction.

The displacement (rippling) of antimony incorporated into the copper surface is also consistent with the picture of the larger antimony atoms substituting for copper however the STM data illustrates that the hexagonal arrangement of the Cu substrate atoms have been perturbed (figure 5.8 (c)). The $(\sqrt{3} \times \sqrt{3})R30°$ structure is considered to be the more thermodynamically stable configuration given that it can be seen from both dissolution and segregation, thus the long range structure in figure 5.8 (a) observed after annealing at high temperatures might be due to kinetic limitations at the surface.

Many alloys do not have long–range chemically ordered phases at all or their order–disorder transition temperature is too low, so that the mobility of the atoms is too small to form long range ordered structure during annealing and cooling down. This is mainly a problem of the bulk, but it can also occur in the surface layer, where diffusion is usually much faster. The STM data shows that $(\sqrt{3} \times \sqrt{3})R30°$ is still present at this elevated temperature together with a (2×1) structure as seen on the proposed schematic of figure 5.8 (d) in real space. Similar structures showing $(\sqrt{3} \times \sqrt{3})R30° + (2×1)$ have been proposed before for superlattices on substrates with hexagonal lattice with their corresponding LEED pattern similar to the schematic in figure 5.8 (d) [22].

Previous studies by Comin et al. [6] reported $(2\sqrt{3} \times \sqrt{3})R30°$–Sb patterns for Te covered Cu(111) and Cu(100) at coverages of 0.33 and 0.25 ML. Given that Te is comparable in size to Sb, it is highly appealing to expect a similar behavior when Sb is grown on Cu(111) at monolayer coverage. It is highly probable that at this annealing temperature, the Cu atoms desorbed from the surface or alternatively diffused to the bulk to leave a fewer copper atoms than in the conventional $(\sqrt{3} \times \sqrt{3})R30°$–Sb structure. Other explanations will be the nearly substitutional Sb dimerization previously displaces Cu adatom decoration, or the combination of these factors. The STM measurements are not enough to distinguish between these possibilities, thus complementary techniques are required for the understanding of the interesting overlayer structure.

5.6 Segregation of Sb on Cu(111)

To study the segregation behavior of Sb on Cu(111), the single crystal was doped by ~ 0.02 at. % of Sb at the backside. The sample was annealed for 3 weeks at 900°C under argon atmosphere. The sample was introduced to the VT–STM UHV system and subjected to the cleaning cycles as in **section 5.2**. Surface concentration of the Sb atoms segregating to the surface of Cu was monitored by Auger electron spectroscopy while the sample temperature was increased linearly from 300 K to 960 K. The Auger peak–to–peak height was acquired using the AugerProfiler program. The AES peaks for Sb (488–462 eV) and Cu (910–925 eV) were recorded at a beam current of 10 µA with a beam voltage of 3 keV. To rule out the presence of impurities during the Auger run, a full AES spectra was taken before and after each run.

The AES peak–to–peak heights were converted to surface concentration and plotted as a function of sample temperature in figure 5.9. The fit to experimental data was normalized to 33 at% maximum surface Sb coverage. Two simulations using the modified Darken model were considered. The first simulation was with Sb bulk concentration of 0.2 at% and the second with Sb bulk of 0.3 at%. The segregation parameters obtained for the fits are summarized in the inset of figure 5.9. From the graph it is clear that the higher Sb bulk concentration is the better fit. The fits were a good indication that the measured data has a typical segregation profile.

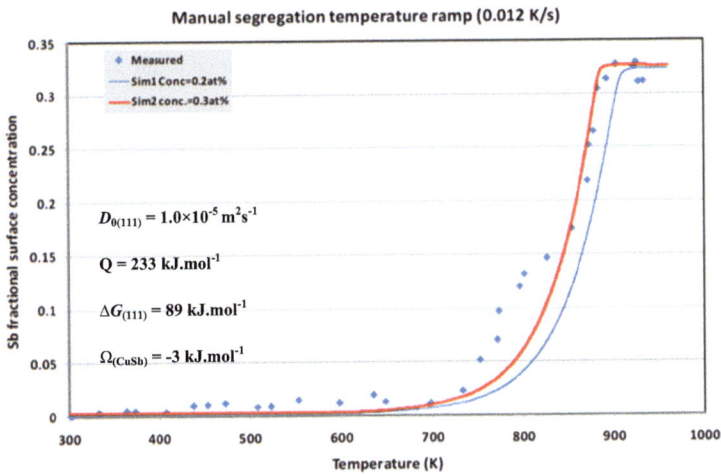

Figure 5.9. The experimental segregation results for the Cu(111) surfaces of Cu (0.02 % Sb). The data were obtained by increasing the crystal temperature linearly at a rate of 0.012 Ks^{-1} while the Sb concentration at the surface was measured as a function of temperature. The simulation with 0.3 at. % (red curve) was a best fit.

The maximum concentrations calculated by AES are in agreement with the ($\sqrt{3} \times \sqrt{3}$) $R30°$ superstructure that was observed by LEED, which indicates a 33 at.% coverage of the surface. The same behavior but in a symmetric way is also observed for dissolution kinetics. Thus, the segregation studies gave a very good indication of the surface coverage at different temperatures during STM measurements on the dissolution CuSb system. The AES data show a decrease in Sb concentration towards the end of the segregation run which is collaborated by the reverting of the LEED pattern from a high temperature structure of the metastable ($\sqrt{3} \times \sqrt{3}$) $R30°$ + (2×2) to the more energetically favourable ($\sqrt{3} \times \sqrt{3}$) $R30°$ structure. While the interaction between molecules adsorbed on surfaces in general is dominated by Van der Waals or electrostatic forces and hence is attractive, however in this study a repulsion interaction (Ω_{CuSb} = -3 kJmol^{-1}) was found to be dominant between the Cu and Sb atoms. The effect is due to an exchange of electric charge between the Sb and Cu substrate atoms and a minimization of the interface potential of the CuSb system.

5.7 Summary

In spite of the immiscibility and large atomic size mismatch (15%) in Cu–Sb system, surface alloy formation at the sub–monolayer region was observed even at ~360°C. The STM experimental data support a structural model for the Cu(111)($\sqrt{3} \times \sqrt{3}$) $R30°$–Sb structure in which Sb atoms displace up to 1/3 of the first–layer of Cu atoms at ~400°C. The Sb atoms are displaced outward with respect to the first–layer Cu atoms. This structural phase was obtained for both dissolution and segregation samples. Annealing to elevated temperatures transforms the Cu(111)($\sqrt{3} \times \sqrt{3}$) $R30°$–Sb structure to the less energetically favorable Cu(111)($2\sqrt{3} \times 2\sqrt{3}$) $R30°$–Sb and ($2\sqrt{3} \times \sqrt{3}$) $R30°$–Sb metastable structures. This structural transition has been reported on Cu–Te system but not in Cu–Sb systems. A complete summary of the structural transition followed as a function of increasing temperature and surface Sb concentration on the Cu–Sb system is tabulated in table 1.

Additional measurements using complementary techniques are required to fully characterize the long range properties and to fully understand the driving force responsible for the occurrence of the new structures. The fact that there is rippling observed in all the reported surface structures is a reminder that redistribution of electronic charge at the surface creates a relaxation which also plays a pivotal role in producing the rippling on metal on metal surfaces.

Temp °C	Structural phase	Concentration %	Adsorption sites	Techniques	References
400	$(\sqrt{3} \times \sqrt{3})R30°$	30	HCP	STM,LEED,AES	[7-9,18]
600	$(2\sqrt{3} \times 2\sqrt{3})R30°$	43	FCC hollow	STM,AES,LEED	[3,20,21,28]
700	$(2\sqrt{3} \times \sqrt{3})R30°$	25	HCP	STM,AES	[6,22]

Table 1. Summary of the acquired experimental data on Cu(111)-Sb system as a function of annealing temperature.

The higher relative corrugation of the topmost Sb atoms compared to the Cu atoms observed at 700°C could be evidence for a high local density of states on Sb, possibly resulting from the occurrence of Sb–Cu bonds. Accordingly a superposition of the Cu and Sb atoms in the dimer is imaged as single protrusion with the STM. Despite the complexity of the observed STM data obtained at elevated temperatures, for the reported structural phases, the data were best described by a model involving an ordered p(2×2), p(2×1) overlayer structure superimposed on the $(\sqrt{3} \times \sqrt{3})R30°$–Sb surface (figure 5.10). The resolution of the data collected on the segregation profile was not enough to fully conclude on the presence of the high temperature structures observed determined on the segregation sample.

5.8 References

[1] I. Gierz, B. Stadtmüller, J.Vuorinen, M. Lindroos, F. Meier, J. H Dil, K. Kern, and C. R. Ast, Phys. Rev. B 81, 245430 (2010)

[2] S. A. de Vries, W. J. Huisman, P. Goedtkindt, M. J. Zwanenburg, S. L. Bennett, I. K. Robinson, E. Vlieg, Surf. Sci. 414 (1998) 159–169

[3] T. C. Q. Noakes, D. A. Hutt, C. F. McConville, D. P. Woodruff, Surf. Sci.372 (1997) 117–131

[4] H. Giordano, B. Aufray, Surf. Sci. 307–309 (1994) 816–820

[5] C. R. Ast, J. Henk, A. Ernst, L. Moreschini, M. C. Falub, D Pacile, P. Bruno, K. Kern and M. Grioni, Phys. Rev. Lett.104, 066802 (2010)

[6] F. Comin, P. H. Citrin, P. Eisenberger, J. E. Rowe, Phys. Rev. B 26 (1982) 7060

[7] H. Giordano, O. Alem, and B. Aufray, Scr. Metall. Matter. 28, 257 (19993)

[8] P. Bailey, T. C. Q. Noakes, D. P. Woodruff, Surf. Sci. 426 (1999) 358

[9] D. P. Woodruff and J. Robinson, J. Phys. :Condens. Matter 12 (2000) 7699

[10] I. M. McLeod, V. R. Dhanak, M. Lahti, A. Matilainen, K. Pussi and K. H. L. Zhang, J. Phys.: Condens. Matter 23 (2011) 265006

[11] D. P. Woodruff, M. A. Munoz–Marquez and R. E. Tanner, Curr. Appl. Phys. 3, 19 (2003)

[12] S. Oppo, V. Fiorentini, M. Scheffler, PRL. 71,1 (1993)

[13] I. Meunier, J. M. Gay, L. Lapena, B. Aufray, H. Oughaddou, E. Landemark, G. Falkenberg, L. Lottermoser and R. L. Johnson, Surf. Sci. 422, 42 (1999)

[14] G. Hormandinger, Phys. Rev. B, 49, 19 (1994), F. Besenbacher, Rep. Prog. Phys. 59 (1996) 1737–1802

[15] T.B. Massalski: in Binary Alloy Phase Diagrams, T.B. Massalski, ed., ASM International, Materials Park, OH, 1990, **pp. 3320-21**

[16] Landolt–Bornstein, Group 3, Vol. 6, Springer, Berlin (1971)

[17] K. Umezawa, H. Takaoka, S. Hirayama, S. Nakanishi, W.M. Gibson, Curr. Appl. Phys. 3 (2003) 71

[18] B. Aufray, H. Giordano, D. N. Siedman, Surf. Sci. 447 (1996) 180–186

[19] A. Tetsuya, Surf. Sci.Rep. 61, 6 (2006) 283–302

[20] E. A. Soares, C. Bittencourt, V. B. Nascimento and V. E. de Carvalho, C. M. C. de Castilho, C. F. McConville, A. V. de Carvalho, D. P. Woodruff, Phys. Rev. Lett. B, 61, 20 (2000)

[21] R. L. Gerlach and T. N. Rhodin, Surf. Sci. 17, 32 (1969)

[22] K. Oura, V. G. Lifshits, A. A. Saranin, A. V. Zatov, M. Katayama, Surface Science an Introduction, Springer, Ney York (2003)

[23] J. J. Terblans, W. J. Erasmus, E. C. Viljoen and J. du Plessis, Surf. Interface Anal. 28 (1999) 70–72

[24] K. D. Sattler, Clusters and Fullerenes, CRC Press (2009)

[25] B. Kaiser, B. Stegemann, European Journal Of Chemical Physics And Physical Chemistry, 5, 1 (2004) 37–42

[26] T. M. Bernhardt, B. Stegemann, B. Kaiser, and K. Rademann, Angew. Chem. Int. 42, 2 (2003)

[27] P. Varga and M. Schmid, Appl. Surf. Sci. 141 (1999) 287

[28] H. Wider, V. Gimple, W. Evenson, G. Schatz, J. Jaworski, J. Prokop and M. Marszałek, J. Phys.: Condens. Matter 15 (2003) 1909–1919

[29] T. D. Daff, I. Saadoune, I. Lisiecki, N. H. de Leeuw, Surf. Sci. 603 (2009)

[30] H. Häkkinen and M. Manninen, Physical Review B 46, 3 (1992) 1725

[31] R. Ferrando, R. Spadacini, and G. E. Tommei, Phys. Rev. B, 45, 1 (1992)

[32] H. Hakkinen and Uzi Landman, Phys. Rev. Lett, 71, 7 (1993)

CHAPTER 6

SCANNING TUNNELING SPECTROSCOPY (STS)

6.1 Introduction

The electrons in surface states of materials are confined to the surface normal between the vacuum barrier and a crystal surface. Surface state wave functions decay exponentially into the solid with a decay length dependent upon the magnitude of the band gap. The states are separated in momentum from the continuum of bulk levels which also occur at a metal surface. In the presence of an impurity, electron states will scatter off the impurity incurring a phase, and for the continuum states these phases will largely interfere and so leave the primary variation in the LDOS that due to the surface state. Surface states which occur near the centre of the surface Brillouin zone (BZ) decay least rapidly into the vacuum, and hence provide the dominant contribution to the LDOS well outside the metal surface-precisely the regime of operation of the STM which can measure (approximately) the LDOS.

The capability of the STM to identifying the chemical nature of an adsorbed species by its apparent height with respect to the sample surface motivated intense efforts in studying the electronic structure of both Cu and Sb species present at the sample surface using the Scanning Tunneling Spectroscopy (STS) technique.

The power of the STM/STS to measure the electronic structure with sub–atomic spatial resolution and the reliability of these measurements depends on the condition of the tunnel junction. When atomic resolution is achieved, the tip DOS is usually highly structured. This fact must be considered carefully in the interpretation of the simultaneously acquired tunneling spectra. During CITS, the data can be overlaid with a grid indicating the positions of the individual *I–V* acquisitions and the individual *I–V* curves can be selected and presented on an *I–V* curve window. Tunneling spectroscopy is able to detect valence states within a few *eV* of the Fermi level and to provide chemical images.

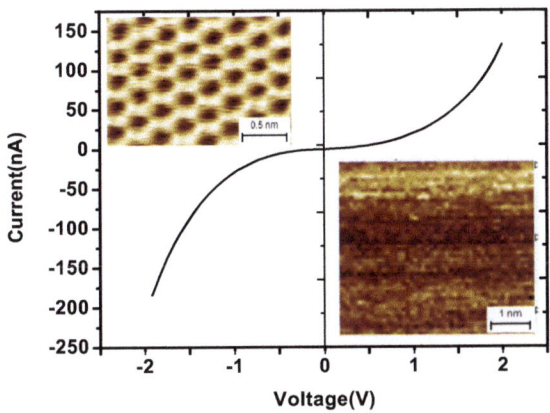

Figure 6. average *I-V* curve of HOPG The top inset in figure 6 is a topography STM image of HOPG while the bottom inset is the constant current image tunneling spectroscopic (CITS) micrograph taken with a bias of 0.3 V and 2 nA tunneling conditions.

STS thus provides essential information about chemically–active valence orbitals, such as the HOMO and LUMO, i.e., the highest occupied and lowest unoccupied molecular orbitals, respectively. As in the constant–current mode, the image is scanned pixel per pixel, but additionally, a current–voltage curve is taken at every point of the image. To test the spectral resolution of the STS mode, measurements were conducted on pristine clean HOPG surface (figure 6). All acquired tunneling spectra represent an average of all curves at the surface to produce a single *I–V* curve. HOPG has a gap in the momentum space, i.e., the waves are out of phase, thus depresses the conductance around the Fermi level and consequently bends the LDOS at low bias voltages and thus the shape of the spectra in figure 6. The top inset in figure 6 is a topography STM image of HOPG while the bottom inset is the constant current image tunneling spectroscopic (CITS) micrograph taken with a bias of 0.3 V and 2 nA tunneling conditions.

6.2 STS of Cu(111) at room temperature

STS experiments were carried out on a surface with well–defined electronic structure Cu(111), with the aim to characterize the electronic states of the sample surface. The basic and theoretical concepts which underlie the STS technique are outlined in **section 3.5**. The I–V curves spectra were extracted from the current imaging tunneling spectroscopy (CITS) data (figure 6.1 (a)). The lighter spots represent larger field emission current from Cu atoms.

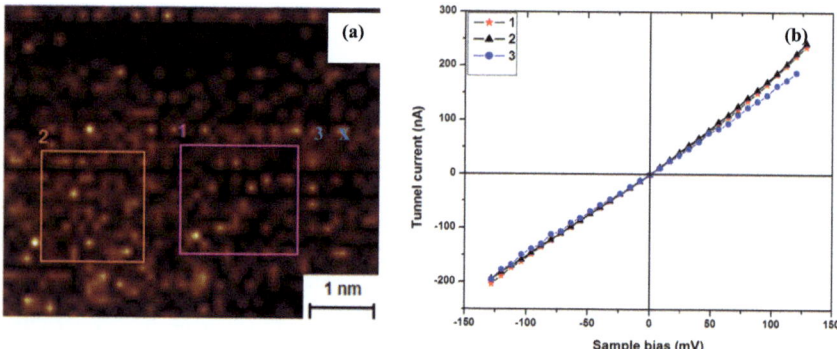

Figure 6.1. (a) Spatially resolved constant current image tunneling spectroscopic (CITS) micrograph of Cu(111) obtained at fixed bias voltage of 2.5 mV. (b) I–V curves spectra corresponding to the areas marked in (a) showing metallic behavior of the copper sample at room temperature.

The *I–V* data represents an average of all curves on the surface to produce a single *I–V* curve representative of the sample surface (figure 6.1 (b)). The images, spectra, and *I–V* curves were all reproducible. The measured *I–V* curves depicts an ideal metallic behavior close to the Fermi level (figure 6.1 (a)). The details of the spectra, such as the slight asymmetry or the distinct peak seen in other regions marked (3) in (a), can safely be attributed to sample properties such as the presence of defects at the sample surface. The curves in (b) are taken along the *I–V* map in (a). This provides the significant possibility of distinguishing surface atoms of different chemical natures accessible by STM at the sample surface.

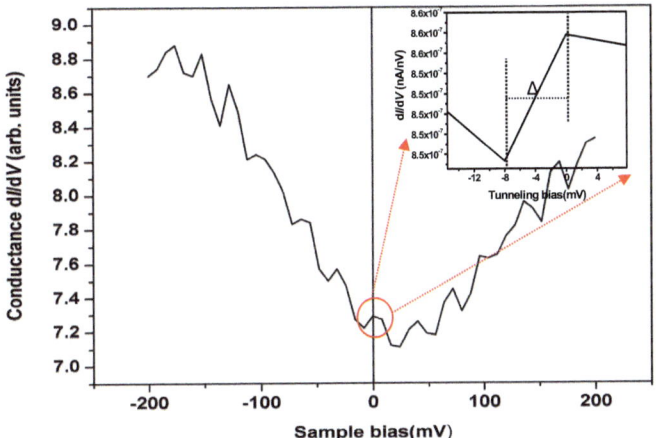

Figure 6.2. Average conductance spectra of the Cu(111) sample surface at room temperature which is the first derivative of the *I–V* map in 6.1 (a). The inset iluustrates unpertubed surface state near E_F showing the step-like onset near -8.0 mV.

The first derivative dI/dV can be obtained from the measured I–V curves by numerical differentiation. Although a slight variation in the shape of these spectra is sometimes seen in figure 6.2, the basic features are very reproducibly observed. Area–averaged conductance spectra show the well-known sharp, step–like increase in the LDOS which is typical for a 2D electron system (figure 6.2). Electronically speaking, most metals surfaces resemble a free electron gas more than orbitals with relatively well–defined energies. The inset in figure 6.2 depicts a sharp feature at -8 meV which can be directly interpreted in terms of the surface-state band edge, i.e. for $E_{\bar{\Gamma}}$=-8 mV Cu(111). The obtained dI/dV spectrum is essentially identical to the one obtained by Vázquez de Parga et al. [1] on Cu(111) at room temperature which exhibits only a peak with the onset at 0.4 eV. The shape of their conductance spectra originated from I-V curves with a metallic behavior close to the Fermi level which is characteristic of the Shockley type surface states on Cu(111) (figure 6.2)). The drastic feature, which is the sudden drop in dI/dV at ~ 8.0 meV can be explained as the sudden decrease in the surface LDOS at energies more than 8 meV below E_F which reflects the onset of tunneling from the surface state band. One source of error in the interpretation of STS data comes from tip-height variations, in this study, this error can be ignore since the feedback loop is switch off during data acquisition. In a similar study by Everson et al. [2] on Au(111) at room temperature, the dI/dV curves were explained in terms of surface states which arose on the (111) surfaces of noble metals as a result of the gap that exists along Γ-L line in their bulk band structure [3,4]. This kind of spectrum is typical of a large number of spectrums which were recorded under a variety of tunneling conditions at room temperature. What is more consistent from spectrum to spectrum is the width of the onset, which forms the basis of the analysis above. To quantify this width, which is denoted Δ, the geometrical definition is illustrated by the inset, extrapolating the slope at the midpoint of the rise to the continuation of the conductance above and below the onset. The width Δ of the onset is related to the surface-state life time $\tau = \hbar/\Delta$ [5-8]. Perhaps the most spectacular applications of this spectroscopic measurements is that, the feature in STS spectra would be the only manifestation of surface states, if the crystal surface were perfect, i.e. if there were no structural imperfections such as step edges and adatoms. However, it is well known that real surfaces possesses kinks, steps, vacancies, etc, thus the periodic potential experienced by the surface-state electrons can be considerably altered in close proximity to defects, and this naturally leads to scattering by foreign surface and sub-surface atoms.

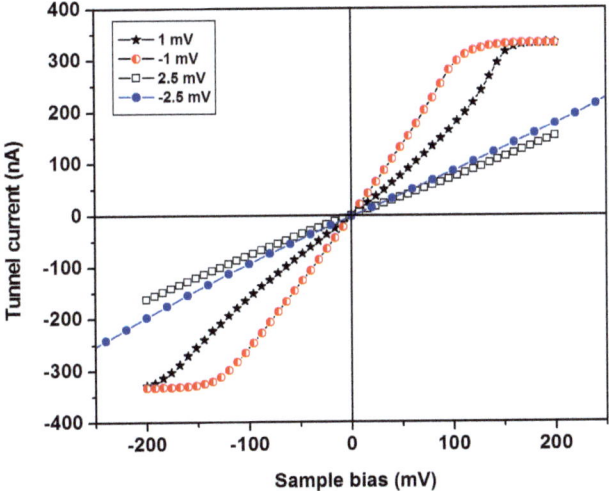

Figure 6.3. *I–V* curves of Cu (111) sample showing variations of the curve shape with applied bias voltage at different polarities.

While these scattering processes can affect Δ of the surface state onset, a quantitative analysis of these effects in terms of Δ is desirable. This property is of interest when studying the electronic properties of segregated controlled amount of impurity atoms (eg., Sb) at the sample surface of low index planes of noble metals.

To aid in the interpretation of the STM/STS acquired data of the Cu sample surface, it is desirable to carry out systematic imaging varying the set point current and polarity thereby changing tip–sample separations. The positions of the peaks shift with applied voltage are in figure 6.3. For comparison, I–V spectra taken with sample bias of ± 1 mV show a saturation threshold at 130 mV as seen in figure 6.3. It was observed that the I–V curves measured over small voltage (± 100 mV) intervals were linear exhibiting metallic behavior . By increasing the voltage range, the nonlinearity first increased slightly, and when the voltage range exceeded the value of ($E_{con\ band}$–E_F) for reverse and ($E_{vale\ band}$–E_F) for forward voltages, the nonlinearity increased dramatically, as shown by plots taken at ± 1mV. (bias polarity determines whether unoccupied or occupied sample electronic states are probed). The most interesting and surprising result obtained from the I–V curves shown in figure 6.3 is that the tunnelling current strongly saturates at ± 1mV sample voltages as compared to ± 2.5 mV. This saturation is direct evidence that the flow of the tunnel current is limited by some sort of charge transport mechanism in the sample. By varying the amount of the applied bias voltage, one can select the electronic states that contribute to the tunneling current and, in principle, measure the local electronic density of states. For instance, the current increases strongly if the applied bias voltage allows the onset of tunneling into a maximum of the unoccupied sample electronic density of states. At a negative sample bias, mainly the sample's empty states are probed with negligible influence of the tip's occupied states. On the other hand, tunneling from the sample to the tip is much more sensitive to the electronic structure of the tip's empty states which often influences the spectroscopic STM studies of the sample's occupied states.

A series of normalized differential conductance measurements were performed for all the curves reported in figure 6.3 and a typical spectrum is shown in figure 6.4. The dependence of the measured spectroscopic data on the value of the tunneling conductance I–V can be compensated by normalizing the differential conductance dI/dV to the total conductance I–V showed that the normalized quantity reflects the electronic density of states reasonably well by minimizing the influence of the tip–sample separation.

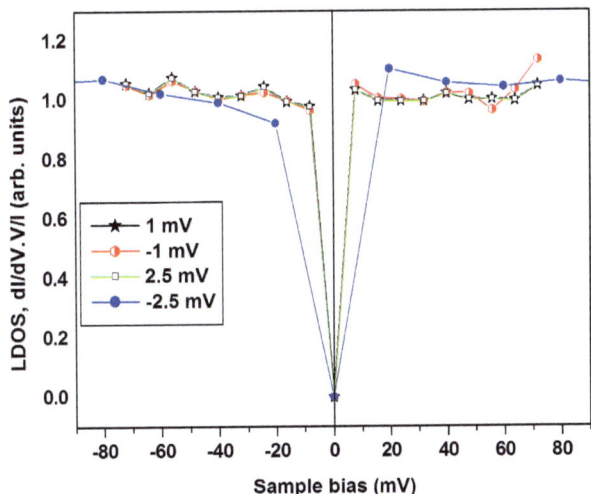

Figure 6.4. Series of normalized differential conductance spectra illustrating the DOS of the Cu (111) sample surface.

Depending on the crystalline structure and morphology of the tip, the differential conductance spectra, either show only the surface state of the bare Cu surface or are dominated by intense peaks due to electronic tip resonances. Metals do not possess a gap between the valence and conduction band, and thus the variations in DOS are relatively small in Cu (figure 6.4).

6.3 Summary

The electronic properties of the Cu(111) surface were studied by CITS at room temperature over different bias voltages. The electronic structure of vicinal Cu surfaces shows a two-dimensional behavior for low energies just above the onset of the surface state. In the case of noble metals, a steplike feature in dI/dV curves is explained in terms of a two-dimensional surface-state band. Ultimately, regardless of the particular materials system chosen, the research proposed in this thesis can be applied to enhance the knowledge in the field of nano sciences. Furthermore, if these avenues of research are pursued, it would enable a full characterization of the nanostructures that might be used as the components and interconnects in future nanoelectronic devices. The two different approaches: differential conductivity dI/dV or the normalized differential conductivity $(dI/dV)I/V$ were used to interpret STS data. The I-V characteristic measurements of the Cu(111)-Sb system at various temperatures which present new challenges for investigation of 2DEG system at atomic scale are ongoing.

6.4 References

[1] A. L. Vázquez de Parga, O. S. Hernán, and R. Miranda, Phys. Rev. Lett. 80, 2 (1998)

[2] M. P. Everson, R. C. Jaklevic, and W. Shen, J. Vac. Sci. Technol. A8, (1990) 3662–3665,

[3] W. Shockley, Phys. Rev. 56, (1939) 317–323

[4] F. Z. Forstmann, Phys. 235, (1970) 69–74

[5] J. T. Li, W. D. Schneider, R. Berndt, O. R. Bryant, S. Crampin, Phys. Rev. Lett. 81, 4464 (1998)

[6] J. Kliewer, R. Berndt, E. V. Chulkov, V. M. Silkin, P. M. Echenique, S. Crampin, Science 288, 1399 (2000).

[7] M. Pivetta, F. Silly, F. Patthey, J.P. Pelz, W.D. Schneider, Phys. Rev. B 67, 193402, (2003)

[8] M. Ternes, M. Pivetta, F. Patthey, W. D. Schneider, Progress in Surface Science, 85 (2010) 1–27

CHAPTER 7

SUMMARY AND CONCLUSION

High spatial resolution STM images were attained on HOPG, Si(111), and Cu(111) surfaces and the acquired data was used for the VT-STM calibration purposes. The acquired hexagonal crystal structures of the cleaved HOPG together with the 0.246 ± 0.02 nm C–C distances were comparable to literature values. Other than the conventional honeycomb and triangular structures of HOPG, superlattices with periodicity of ~ 3.45 ± 0.02 nm called Moiré patterns were observed in atomic details. The complex Si(111)-7×7 reconstruction was obtained and the observations of surface faulted and unfaulted half unit cells of dimensions 2.76 ± 0.01 nm were in good agreement with reported theoretical and experimental data.

After repeated cycles of sputtering and annealing in UHV, an atomically clean and well ordered Cu(111) sample was revealed in atomic detail by STM. The Cu–Cu atomic spacing was measured to be ~ 0.257 ± 0.01 nm on average and the angle between adjacent rows of Cu atoms was found to be 60° which is characteristic of a Cu(111) crystal surface.

The *in situ* STM system was further used in conjunction with LEED and AES to grow and characterise monolayer Sb coverages grown from a Knudsen effusion cell on surfaces of Cu(111) in UHV. The total concentration of Sb atoms at the sample surface was calculated from STM data after growth and was determined to be ~ 0.43 ML on average. After growth Sb atoms were randomly arranged on the Cu(111) surface. The Cu–Sb surface alloy's evolution was followed as a function of annealing temperature from 300°C up to 700°C. Although Sb is much larger than Cu, this large atomic size mismatch does not impede surface alloy formation even at 300°C (figure 7.1). At this annealing temperature, the initial stages of dissolution of Sb onto the Cu surface is evident from the re-arrangement of Sb atoms at the surface from random distribution to recognizable patches of hexagonal arrangement. Highly crystalline alloy overlayers having distinct structural phases ranging from the conventional and energetically most favourable ($\sqrt{3} \times \sqrt{3}$) $R30°$–Sb at 400°C, and metastable ($\sqrt{3} \times \sqrt{3}$) $R30°$–Sb + p(2×2) at 600°C, and ($\sqrt{3} \times \sqrt{3}$) $R30°$–Sb + p(2×1) structure at 700°C were obtained for the first time (figure 7.1).

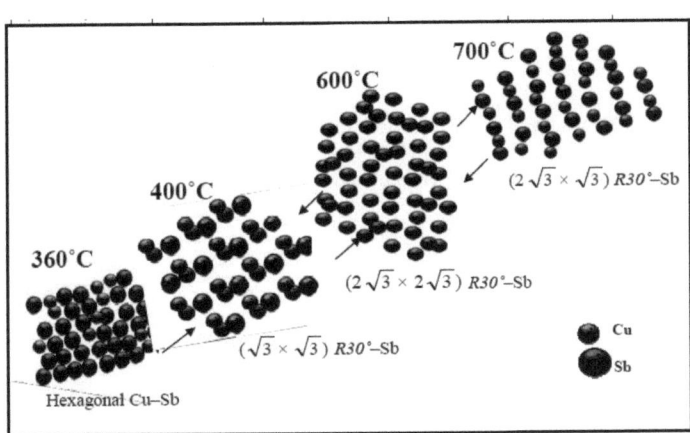

Figure 7.1. Schematic models of the Cu–Sb and their corresponding annealing temperatures.

The coverage associated with the ($\sqrt{3} \times \sqrt{3}$) $R30°$–Sb phase was found to be 0.3 ML and clearly favours a structural model based on Sb atoms occupying substitutional rather than overlayer sites within the top Cu(111) layer substituting one–third of the Cu atoms. At the monolayer Sb coverage, STM results have shown that the Sb–induced compressive stress gets so high that it is relieved by segregation to the surface by some of the Sb atoms that were formerly alloyed into the surface layer resulting in the reported metastable structural phases at various temperatures. Furthermore, the STM micrographs also show that the six Cu atoms surrounding each Sb atom, in the (111) plane, are strongly perturbed, since they appear in the atomically resolved images to be vertically and laterally displaced from their expected positions resulting in a broken surface symmetry which is a desirable effect in spintronics application. The acquired topographic STM data was consolidated by segregation studies utilising AES technique on the surface alloy at different temperatures. LEED, AES and STM studies were used to distinguish between atomic positions of Cu and Sb in the complex metastable $(2\sqrt{3} \times 2\sqrt{3})$ $R30°$–Sb structural phase and a possible structural model for the ordered Sb surface phase was proposed. This surface alloy structural phase was only stable in the temperature range 240–250°C after which the Sb surface concentration dropped from 0.34 ML to 0.3 ML which is a critical coverage to induce the $(\sqrt{3} \times \sqrt{3})$ $R30°$–Sb superstructure. For the metastable structural phase obtained at 600°C and 700°C, the data are best described by a model involving an ordered p(2×2) and p(2×1)–Sb overlayer (periodicity relative to the substrate) superimposed on the $(\sqrt{3} \times \sqrt{3})$ $R30°$ – Sb surface alloy, respectively. These results demonstrate that the balance between the interface energy and the elastic strain energy determines the stabilization of different interface structures on the atomic scale. In particular, a continuous change in lattice parameters with increasing temperature was found for the long-range ordered structures, indicating a substrate-mediated repulsive intermolecular interaction as confirmed by AES studies. Not only was the registry of the top layer atoms, but also their out–of–plane relaxations were determined. The acquired surface structures on the Cu–Sb system as a function of annealing temperature are summarized by the schematic model on figure 7.1.

Surface structures	Annealing Temperature (°C)	Cu–Cu (nm)	Sb–Sb (nm)	Roughness (nm)	Sb concentration (%)
Cu(111)	27	0.257±0.02	-	1.613	-
Semi hexagonal	360	0.461±0.02	0.587±0.03	2.031	43
($\sqrt{3} \times \sqrt{3}$) $R30°$–Sb	400	0.626±0.01	0.826±0.001	1.706	30
($2\sqrt{3} \times 2\sqrt{3}$) $R30°$–Sb	600	0.286±0.001	0.816±0.002	2.054	43
($2\sqrt{3} \times \sqrt{3}$) $R30°$–Sb	700	0.757±0.002	0.817±0.001	4.013	25

Table 2. Summary of the acquired STM data on the Cu–Sb system.

The surface roughness was measured as a function of annealing temperature (Table 2). There data show an increase in roughness after growth of Sb at the surface of copper which latter decrease when the equilibrium structure is obtained at 400°C. At equilibrium, the surface atoms rearrange to a more energetically stable configuration and thus lower the surface energy of the system. There after there is an increase in sample roughness with increasing temperature which is expected as the diffusion of surface atoms increases with increasing temperature and also evident from the increase in spacing between atoms (both Cu–C and Sb–Sb) at high temperatures. Therefore both figure 7.1 and Table 2 gives a inclusive description of the behavior of surface atoms at the Cu–Sb system as a function of annealing temperature.

The spectroscopic information with an unprecedented spatial resolution ultimately at the atomic level, such as site–specific information is generally not provided by area–

integrating surface analytical techniques. By performing atomically resolved current imaging tunneling spectroscopy (CITS) measurements, the atomic origins of the various electronic states of the Cu(111) were directly determined. The acquired data show metallic behavior of the Cu sample at room temperature. The STM conductivity has been acquired for Cu(111) surface using both the direct measurements from conductance measurements and DOS maps at finite voltages and in the presence of surface states. The technique of STS is still in its infancy, thus the ability to pinpoint or relate the topographic data on a surface with different atomic species to the spectroscopy signal and consequently the ultimate goal of elemental identification has proven be a challenging task. Nevertheless, the possibility of measuring the local down to atomic resolution electronic density of states of Cu(111) surface using both the conductance and differential conductance maps has been demonstrated. This approach should be applicable to studies of perturbations of both bulk and surface metallic band structure by adsorbates. Improvement in the resolution of the STS technique is highly desirable.

The consequences of surface diffusion are easily seen in images taken by a scanning tunneling microscope (STM), but its mechanism can hardly be determined because of the limited time resolution of the STM which is of macroscopic dimensions. Thus, consistent efforts should be made in further improvements to make the STM/STS more sensitive, robust and reliable, obviously this requires developing a better understanding of surface processes such as self-assembled monolayers on metallic or semiconducting surfaces.

The successful combination of all the techniques described (STM, STS, LEED, AES) presented crucial information on understanding interactions at the atomic scale, highly relevant in surface science and sub fields. The evolution of scanning probe based science and technology is thus of relevance in the excursion into the nano world fuelled with exciting possibilities. These studies provide part of the basic underpinning understanding that is necessary for surface alloys to be fully exploited for their functional properties such as temperature sensing and catalytic activities. For further application of the surface alloys in research areas such as electronics and other technological applications, a thorough understanding of dopant behavior in these materials is essential. That being said, this study has showed in details that it is possible to engineer or acquire particular arrangement of atomic species on a substrate occupying specific sites as driven by the systematic thermodynamics of the sample surface.

FUTURE PROSPECTS

The studies presented in this thesis, while they have advanced the knowledge in the field of nanoscale surface science growth and characterization, have left ample room for future research. What follows in this chapter are directions identified for future research.

- It would be of considerable interest to have the benefit of theoretical studies which might cast some light on the underlying driving force for the surface reconstructions regardless of the materials system involved and structural phases.

- Investigate and compare the diffusivities, segregation energies and activation energies obtained by Auger spectroscopy with those obtained by scanning tunneling spectroscopy through IV curves and discuss their relation.

- Conduct STS measurements in conjunction with the segregation studies to determine the change in electronic properties on atomic scale of the surface alloys as the Sb concentration increases with increasing temperature which will ultimately results in specific atomic material identification. Such studies will also give insight on the perturbation of the local DOS of the Cu surface by the presence of controlled impurities, in this case Sb atoms.

PUBLICATIONS

G. F. Ndlovu, W. D. Roos, K. T. Hillie, B. W. Mwakikunga, Z. M. Wang, J. K. O. Asante, M. G. Mashapa, C. J. Jafta, Epitaxial deposition of silver ultrafine nano–clusters on defect–free surfaces of HOPG–derived few–layer graphene in a UHV multi–chamber by *in–situ* STM, XPS and *ab* initio calculations,(Accepted, *Nanoscale Research Letters*).

Gebhu F Ndlovu, Bonex W Mwakikunga, Aimé Lay-Ekuakille, Wiets D Roos, Joseph K O Asantee, Kenneth T Hillie, Microscopy and Spectroscopy Measurements of Sb on Cu(111) Surfaces: Topography and Electronic Properties (Submitted to IEE).

B. W. Mwakikunga, T. Malwela, K. T. Hillie, G. F. Ndlovu, Towards an electronic nose based on nano–structured transition metal oxides activated by a tuneable UV light, source, Proc. IEEE Sensors, CFP11SEN–CDR, 2011.

G F Ndlovu, J K O Asante, W D Roos, K T Hillie, Investigation of broken symmetry of Sb/Cu(111) surface alloys by VT–STM, Proceedings of the 56th Annual Conference of the South African Institute of Physics, University of South Africa, July 2011

G F Ndlovu, J K O Asante, W D Roos, K T Hillie, The decoration of vicinal copper polycrystalline surfaces by Antimonym, Proceedings of the 56th Annual Conference of the South African Institute of Physics, University of South Africa, July 2011

G. F. Ndlovu, J. K. O. Asante, W.D. Roos, B.W. Mwakikunga, K.T. Hillie, Showcasing the unprecedented resolution of the variable temperature scanning tunneling microscopy, Proceedings of the 49th Annual Conference of the Microscopy of Southern Africa, Pretoria, 5–9 December 2011

G. F. Ndlovu, J. K. O. Asante, W.D. Roos, B.W. Mwakikunga, K.T. Hillie, Variable temperature scanning tunneling microscopy study of Sb growth on copper (111) surfaces, Proceedings of the 49th Annual Conference of the Microscopy of Southern Africa, Pretoria, 5–9 December 2011

INTERNATIONAL MEETINGS AND CONFERENCES

- B. W. Mwakikunga, T. Malwela, K. T. Hillie, **G. F. Ndlovu,** Towards an electronic nose based on nano–structured transition metal oxides activated by a tuneable UV light source IEE Sensor 2011 Conference, Limerick, Ireland, 28 –31 October, University of Limek

- **G. F. Ndlovu**, J. K. O. Asante, W. D. Roos, K. T. Hillie, Spin Splitting with Surface Alloys – The 4th Korea–South Africa Joint Workshop on Nanotechnology, KIST, Seoul, South Korea, October 2009

- **G. F. Ndlovu**, The study of Quartz Textures in Multiphase rocks using Neutron Diffraction Texture Analysis at the JINR, Dubna Summer Practice for South African Students, Joint Institute of Nuclear Research (JINR), Dubna, Russia, September 2009

LOCAL MEETINGS AND CONFERENCES

- **G F Ndlovu,** J K O Asante, W D Roos, K T Hillie, Variable Temperature Scanning Tunneling Microscopy Study Of Sb Growth On Copper (111) Surfaces, 49th Conference of the Microscopy Society of Southern Africa (MSSA), CSIR, Pretoria, 6–9 December 2011 (**ISBN: 0620350563**)

- **G F Ndlovu,** J K O Asante, W D Roos, K T Hillie, Showcasing The Unprecedented Resolution Of The Variable Temperature Scanning Tunneling Microscopy, 49th

Conference of the Microscopy Society of Southern Africa (MSSA), CSIR, Pretoria, 6–9 December 2011(**ISBN: 0620350563**)

- **G F Ndlovu,** J K O Asante, W D Roos, K T Hillie, Investigation of broken symmetry of Sb/Cu(111) surface alloys by VT–STM, 56th Annual Conference of the South African Institute of Physics, University of South Africa, July 201(**ISBN: 978-1-86888 -688 -3**)

- **G. F. Ndlovu**, J. K. O. Asante, W. D. Roos, K. T. Hillie, The decoration of vicinal copper polycrystalline surfaces by Antimony Ndlovu, J K O Asante, W D Roos, K T Hillie, 56th Annual Conference of the South African Institute of Physics, University of South Africa, July 2011(**ISBN: 978-1-86888 -688 -3**)

- **G. F. Ndlovu**, J. K. O. Asante, W. D. Roos, K. T. Hillie, Atomic resolved Ag nano clusters on HOPG: a VT–STM study, 55th Annual Conference of the South African Institute of Physics, National Laser Centre, October 2010

- **G. F. Ndlovu**, J. K. O. Asante, W. D. Roos, K. T. Hillie, Calibration of the Variable Temperature Scanning Tunneling Microscope for High atomic Resolution, Nano Africa conference, Council for Scientific and Industrial Research, February 2009

- **G. F. Ndlovu**, J. K. O. Asante, W. D. Roos, K. T. Hillie, Variable Temperature (VT) Scanning Tunnelling Microscopy (STM) Studies of Si(111) and HOPG, 54th Annual Conference of the South African Institute of Physics, University of Kwa Zulu Natal, July 2009

- **G. F. Ndlovu**, J. K. O. Asante, W. D. Roos, K. T. Hillie, Spin–Splitting Monolayer Alloys, National Center for Nano–structured Materials Colloquium, CSIR, SA, August 2008

- **G. F. Ndlovu**, 53rd Annual Conference of the South African Institute of Physics, University of Limpompo, July 2008, In attendance.

AWARDS

- **Frank Nabaro** prize for the most outstanding oral presentation in the field of Condensed Matter physics and/ or Materials Science delivered at the annual SAIP conference by a Doctoral student, July 2011, Pretoria

- **Wirsam Tescan** prize for the most exceptional presentation at the 49th annual meeting of the Microscopy Society of Southern Africa (MSSA), CSIR, Pretoria, 6–9 December 2011

- **SMM** prize for the best paper on an innovative microscopy technique at the 49th annual meeting of the Microscopy Society of Southern Africa (MSSA), CSIR, Pretoria, 6–9 December 2011